Geometric Algebra in Multi-dimensional Signal Processing

（多维信号处理的几何代数方法）

Wang Rui（王　瑞）　Cao Wenming（曹文明）

Science Press
Beijing

Wang Rui
School of Communication and Information Engineering, Shanghai University

Cao Wenming
College of Information Engineering, Shenzhen University

Published by Science Press
16 Donghuangchenggen North Street
Beijing 100717, P. R. China

Printed in Beijing

ISBN 978-7-03-060539-9

Preface

Geometric algebra is a powerful and practical framework for the representation and solution of multi-dimensional signal processing problems. We believe it to be eminently suitable to those sub-fields of signal processing in which such issues occur: multimedia, multispectral images, multi-dimensional signal processing, data streaming and computer vision. We wrote this book to explain the basic structure of geometric algebra, and to help the reader become a practical user. We employ various tools to get there.

Chapter 1: We address the l_1-norm minimization problem, which plays an important role in the compressed sensing (CS) theory. While most previous approaches are designed for signals of low dimensions, we present in this chapter an algorithm using geometric algebra (GA) for solving the problem of l_1-norm minimization for multi-dimensional signals with and without noise contamination by converting it to second-order cone programming (SOCP). The algorithm represents the multi-dimensional signal as a multivector in the GA form to process it in a holistic way without losing the correlation among different dimensions of the multi-dimensional signal. Numerical experiments are provided to demonstrate the effectiveness of the proposed algorithm and its robustness to noise. The proposed algorithm can also be used to guide perfect recovery of multi-dimensional signals, and may find potential applications when CS theory meets the multi-dimensional signal processing.

Chapter 2: Singular value decomposition (SVD) in which the singular value vector preserves the correlation of rows or columns in the matrix successfully, plays an increasingly vital role in many image applications. However, current leading SVD algorithms fail to be directly applied in multispectral images. In this chapter, we propose a novel singular value decomposition algorithm based on the geometric algebra (GA), for multispectral image, namely GA-SVD. It is demonstrated by the comparison results that the proposed GA-SVD algorithm can outperform significantly existing state-of-the-art algorithms in the performance of both quantitative and visual aspects, and particularly show its potential as an effective and homogeneous tool in multiple applications.

Chapter 3: Multispectral images can provide more faithful representations for real scenes than the traditional images and improve the performance of image restoration tasks. In this chapter, we propose a novel multivector sparse representation model for multispectral images using geometric algebra (GA), with the truth that geometric algebra is now well used in image processing and it gives a formidable way to represent multispectral images. The proposed model represents a multispectral image as a GA multivector by fully considering the spatial and spectral information, where a GA dictionary learning algorithm is presented using the K-GA-singular value decomposition (K-GASVD) (generalized K-means clustering for GASVD) method. Consequently, with the complete consideration of the relationship between spectral channels in multispectral images, artifacts and blurring effects can be successfully avoided. The experimental results demonstrate that the proposed sparse model surpasses existing methods for multispectral images reconstruction and denoising tasks by capturing correlations between spectral channels thoroughly, and shows its usefulness and effectiveness for multispectral images processing.

Chapter 4: Speeded-up robust features (SURF) can be used for tasks such as object recognition, image registration, classification or 3D reconstruction. However, existing SURF algorithm cannot be directly applied to deal with multispectral images. In this chapter, based on SURF and the theory of geometric algebra (GA), a novel feature extraction algorithm named GA-SURF is proposed for multispectral images. First, a Hessian matrix based on GA is calculated for locating interest points in spatial and spectral space. The box filters based on GA are used to simplify the calculation of Hessian matrix and generate image pyramids. Then, following the procedures of SURF, interest points are located by the image pyramids and described in GA space. Experimental results demonstrate that, compared with other feature extraction algorithms for multispectral images, GA-SURF can be computed much faster and are more robust and distinctive.

Chapter 5: Multi-modal image registration in medical image analysis is very challenging as the appearance of body structures in images from different imaging devices can be very different. In this chapter, we propose a new method (GA-SURF), which incorporates the geometric algebra (GA) into SURF framework, to detect features from images. We model the volumetric data and register the multi-modal medical images using feature spheres formulated in conformal geometric algebra

(CGA). Specifically, we first extract features from medical images using GA-SURF. Second, we construct the feature spheres using the feature points and find the correspondence of feature spheres in the two images using CGA. With that we can register the images based on the correspondence of the feature spheres. The experimental results evaluated by RIRE have shown that our method can register the multi-modal images with high accuracy.The maximum registration error is less than 4mm.

Chapter 6: Spatio-temporal interest points (STIPs) are often extracted for action recognition in videos. But most of the current interest points detection methods are not suitable for multi-channel videos. In this chapter, a new geometric algebra based spatio-temporal interest points (GA-STIP) detection method is proposed for action recognition in multi-channel videos. Specifically, we take advantage of geometric algebra (GA) to represent multi-channel videos, and develop a new method to obtain multi-scale space, then GA based difference of Gaussian are computed to detect spatio-temporal interest points (STIPs). Experimental results show that GA-STIP method can achieve better performance for multi-channel video action recognition.

Chapter 7: Convolutional neural networks (CNNs) have achieved state-of-the-art performance in many computer vision fields. However, traditional CNNs models cannot be directly applied in multi-dimensional signal processing, which inevitably ignore the interrelationship of the multi-dimensional signals. This chapter proposes a novel convolutional neural networks model based on geometric algebra (GA), namely GA-CNNs to deduce an efficient representation of multi-dimensional signals using GA, which processes multi-dimensional signals in a holistic way without losing interrelationship among channels. Specifically, taking advantage of GA, the multi-dimensional signals are represented as GA multivectors to capture the inherent structures and preserve the correlation of those channels. In particular, both the basic structure of our model such as convolution and fully-connected layers and the back-propagation training algorithm, are extended to the GA domain. We evaluate the performance of the proposed GA-CNNs model in classification tasks on different datasets. The experimental results indicate that our GA-CNNs model outperforms the traditional real-valued CNNs with higher classification accuracy and better convergence.

Chapter 8: Sparse representations have been extended to color image processing. However, existing sparse models treat each color image pixel either as a scalar which loses color structures or as a quaternion vector matrix with high computational complexity. In this chapter, we propose a novel sparse representation model for color image that bears multiple channels based on geometric algebra. Firstly, a novel theory of reduced geometric algebra (RGA) is provided, including commutative sparse basis and the geometric operations. Secondly, taking advantage of the RGA theory, the model represents color image with three-channel as a multivector with the spatial and spectral information in RGA space. Thirdly, the dictionary learning algorithm is provided using the K-RGA-based singular value decomposition (K-RGASVD) (generalized K-means clustering for RGASVD) method. The comparison results demonstrate the proposed model can remove the data redundancy and reduce the computational complexity, and can meanwhile effectively preserve the inherent color structures. The result suggests its potential as a homogeneous and efficient tool in various applications of color image analysis.

Chapter 9: This chapter proposes an extended model of multilayer perceptron (MLP) based on reduced geometric algebra (RGA), namely RGA-MLP, for multi-dimensional signals processing, and demonstrates its emerging potential especially in classification tasks. Firstly, multi-dimensional signals are represented as multivectors in RGA space, in which the multiplication rules are commutative. Secondly, the proposed RGA-MLP, in which all neuronal parameters such as inputs, connection weights, activation function and outputs, and also operators are encoded by RGA. Thirdly, we apply the RGA-version back propagation (BP) algorithm for training. Significantly, due to the utilization of RGA with commutative multiplication rules, multi-dimensional signals are processed in a holistic manner which avoids losing relationship of multiple dimensions, and the network can be simplified simultaneously. Finally, experiments on color images classification are conducted and exhibit that the proposed RGA-MLP model achieve faster convergence rate and higher classification accuracy compared with the traditional real-valued MLP model.

Chapter 10: Convolutional neural networks (CNNs) have drawn much attention and become relatively active topics in the field of computer vision and artificial intelligence. However, existing methods cannot be applied directly in multi-dimensional data because they either fail to preserve the interrelationship of different dimensions or generate high learning cost of the networks. This chapter proposes a novel algebraic framework of CNNs based on geometric algebra for multi-dimensional data processing. Firstly, reduced geometric algebra (RGA) with commutative geometric and convolution operations is introduced. Secondly, taking advantage of the

RGA theory, the multi-dimensional data can be represented as a RGA multivector, in which it allows the modeling of a point in multi-dimensional RGA space as a single entity, rather than as a set of multiple data items on which geometrical operations are carried out. Thirdly, neurons, convolution kernels, learning algorithms and all the related computations are defined in RGA space. Fourthly, a novel CNNs, namely RGA-CNNs is proposed for multi-dimensional data under the mathematical framework of RGA. Finally, the experimental results exhibit the effectiveness of the proposed model, which can not only retain the correlation of multiple dimensions but also achieve lower network learning cost than its counterparts.

Acknowledgments

Of those who have helped us develop this work, we especially thank Prof. Xie Weixin, not only for introducing geometric algebra, but also for giving us an early introduction to the geometric algebra model in computer science and signal processing. We also thank Shen Miaomiao, Wang Kaili, Cao Zhongzheng and Shi Yijie for contributions to this book. We gratefully acknowledge the support of the National Natural Science Foundation of China under Grant No. 61771299, 61771322.

Contents

Chapter 1

L_1-norm Minimization for Multi-dimensional Signals Based on Geometric Algebra

1.1 Introduction

The l_1-norm minimization problem plays a vital role in the recently developed compressed sensing (CS) theory[1−3], which is a new method for data acquisition and finds extensive applications in the field of signal and image processing. While CS has conventionally been used to handle real input data, special attention has also been paid to complex input data to address problems such as blind source separation[4], radar signal processing[5], and terahertz imaging[6]. Moreover, the theory and application of quaternion algebra, which has been first proposed by Hamilton[7], has been developed rapidly in recent years (for example, Ref. [8-14]). Ell and Sangwine[8,14] applied the quaternion Fourier transform to color image processing. Jiang and Wei[11] proposed an algorithm for solving the quaternion least squares problem.

$$\min \|Ax - y\|_2, \quad \text{s.t.} \quad Bx = z \tag{1.1}$$

where $A \in \mathbb{Q}^{n\times m}$, $B \in \mathbb{Q}^{n\times m}$, $x \in \mathbb{Q}^{n\times 1}$ and $y \in \mathbb{Q}^{n\times 1}$, $\mathbb{Q}$ denotes the field of quaternion number, and $\|\cdot\|_2$ is the l_2-norm of the quaternion vector x. Zhang et al.[15] proposed an algorithm for recovering the quaternion signals in both noiseless and noise contaminated scenarios by solving the following two l_1-norm minimization problems, respectively.

$$\min \|x\|_1, \quad \text{s.t.} \quad y = Ax \tag{1.2}$$

$$\min \|x\|_1, \quad \text{s.t.} \quad \|Ax - y\|_2 \leqslant \varepsilon \tag{1.3}$$

where $\|\cdot\|_1$ is the l_1-norm of the quaternion vector x and ε denotes the noise penalty level for noise term l_1 in noise measurement $y = Ax + e$. Winter et al.[4] converted the l_1-norm minimization of complex signals to the second-order cone programming (SOCP), which was then solved by SeDuMi software[16]. However, there is few research works focusing on the l_1-norm minimization for multi-dimensional signals.

Fortunately, geometric algebra (GA), also known as Clifford algebra, provides a powerful framework in signal and image processing and it gives a formidable way for multi-dimensional signals[17−23]. GA is a mathematical language and has broad applications in engineering, physics, and computer graphics[24−26]. Ebling and Scheuermann define a Clifford Fourier transform over (multi) vector fields and prove the generalizations of convolution, correlation, and derivative theorems[27]. Wang et al. created a new algorithm called sparse fast Clifford Fourier transform (SFCFT), which can greatly improve computing performance in scalar and vector fields[28]. Li et al. put forward a novel algorithmic framework based on the SIFT for multispectral images[29]. Shen et al.[30] presented a novel theory of reduced geometric algebra (RGA) with commutative multiplication rules and a new vector-valued sparse representation model for color image using RGA.

Based on GA, the l_1-norm minimization can be extended to multi-dimensional signals. In this chapter, we provide a new algorithm using geometric algebra (GA) to solve the problem of l_1-norm minimization for multi-dimensional signals by converting it to second-order cone programming (SOCP), which can be solved by SeDuMi software[16]. The algorithm represents the multi-dimensional signal as a multivector in the GA form to process it in a holistic manner to avoid the loss of correlation among different dimensions of the multi-dimensional signal. The proposed algorithm may find its potential applications when the multi-dimensional signal is needed to be processed based on CS theory.

The rest of this chapter is organized as follows. Section 1.2 reviews the existing methods for solving l_1-norm minimization problem and basis of geometric algebra (GA) briefly. Section 1.3 describes the proposed algorithm for l_1-norm minimization problem for multi-dimensional signals with and without noise. Section 1.4 presents the instance of l_1-norm minimization problem in $\mathbb{G}_2, \mathbb{G}_3$ space. Numerical experiments of 4-dimensional signals and 8-dimensional signals recovery using our proposed algorithm are provided in Section 1.5 for practical guidelines. Section 1.6 concludes the chapter and also points out the direction of further research.

1.2 Related Work

1.2.1 Preliminaries of Geometric Algebra

The geometric algebra (GA)[24–26] is a powerful mathematical tool for geometric calculus, it provides an efficient framework to conduct the computations without using the coordinate information. It completes the constructions and modellings in a coordinate-free manner and also has found wide applications in both physics and computer vision[22,27–29,31].

In this section, we briefly review the basis of GA. Suppose $\mathbb{G}_n$ is n-dimensional space with an orthonormal basis of vectors $\{e_i\}, i = 1, \cdots, n$, which leads to a basis

$$\{1, \{e_i\}, \{e_i e_j\}, \cdots, \{e_1 e_2 \cdots e_n\}\} \tag{1.4}$$

Specifically $\mathbb{G}_n$ can be represented by $\mathbb{G}_{p,q}$, where $n = p + q$ and p, q are the number of vectors with positive square and negative square in the basis of the space, respectively. That is,

$$e_i^2 = \begin{cases} 1, & 1 \leqslant i \leqslant p \\ -1, & p+1 \leqslant i \leqslant n \end{cases} \tag{1.5}$$

In this chapter, we focus on $\mathbb{G}_n$, in which $q = 0$. The geometric product of two basis is anti-commutative, and we have:

$$e_i e_j = e_{ij} = -e_j e_i = -e_{ji}, \quad i, j = 1, \cdots, n, \quad i \neq j \tag{1.6}$$

$$e_i^2 = 1, \quad i = 1, \cdots, n \tag{1.7}$$

$$e_i e_{ij} = e_i e_i e_j = e_j, \quad i, j = 1, \cdots, n, \quad i \neq j \tag{1.8}$$

For vectors w and z in $\mathbb{G}_n$, the geometric product is defined as:

$$wz = w \cdot z + w \wedge z \tag{1.9}$$

where $w \cdot z$ denotes the inner product, $w \wedge z$ denotes the outer product. Since the e_i vectors are orthogonal, then $e_i e_j = e_i \cdot e_j + e_i \wedge e_j = e_i \wedge e_j$, then the geometric product equals to the outer product.

Take $\mathbb{G}_3$ for instance, the orthogonal bases are constructed by vectors with $2^3=8$ grades, which is given by: $\{1,\{e_1,e_2,e_3\},\{e_1e_2,e_1e_3,e_2e_3\},\{e_1e_2e_3\}\}$ and a simpler form: $\{1,e_1,e_2,e_3,e_{12},e_{13},e_{23},e_{123}\}$. For a 2^n-dimensional signal, only n basis is needed. Generally, the outer product of k vectors is called a k-blade. The number k is called the grade of the blade.

The GA gives the complete mathematical relationship between the two vectors. In GA, multivectors, which are the extension of vectors to higher dimensions, are the basic elements. Any multivector $K \in \mathbb{G}_n$ is described by

$$K = E_0 + \sum_{1\leqslant i\leqslant n} E_i(K)e_i + \sum_{1\leqslant i<j\leqslant n} E_{ij}(K)e_{ij} + \cdots + E_{1\cdots n}(K)e_{1\cdots n}, \quad E(K)\in\mathbb{R} \tag{1.10}$$

The l_p-norm of a GA multivector

$$u = \begin{pmatrix} E_0(u) + \sum\limits_{1\leqslant i\leqslant n} E_i(u)e_i + \sum\limits_{1\leqslant i<j\leqslant n} E_{ij}(u)e_{ij} \\ +\cdots + E_{1\cdots n}(a)e_{1\cdots n} \in (\mathbb{G}_n)^M \end{pmatrix}$$

can be represented as

$$\|u\|_p = \left(\sum_{1\leqslant i\leqslant M} |u_i|^p\right)^{1/p}, \quad p=1,2 \tag{1.11}$$

where $E(u)\in\mathbb{R}^M$.

1.2.2 L_1-norm Minimization

The problem of l_1-norm minimization can be handled by the second-order cone programming (SOCP) algorithm, which can be then solved by SeDuMi software[29], and have achieved the state-of-the-art performance on signal processing in real, complex, quaternion and octonion domain. Winter et al.[4] proposed an algorithm to solve the l_1-norm minimization of complex numbers, it worked directly on the complex-valued data by converting it to the second-order cone programming (SOCP) in real domain. Then, this method has been extended to quaternion input data, which was introduced in Ref. [15] by Zhang et al. The l_1-norm minimization problems of quaternion numbers were solved by its equivalent SOCP in real domain, and showed better performance and feasibility of the algorithm. We have previously presented an algorithm for recovering the octonion signals in both noiseless and noise contaminated scenarios by solving an l_1-norm minimization problem[32]. However, there has no generalized framework to deal with the l_1-norm minimization problem for multi-dimensional signals, then in this chapter we mainly aim to find such an efficient algorithm for multi-dimensional signals.

1.3 The Proposed Algorithm

1.3.1 Noiseless Case

Now we will focus on the l_1-norm minimization problems for 2^n-dimensional signals based on GA using SOCP. That is to derive a new algorithm for solving the following optimization problem

$$\min \|u\|_1, \quad \text{s.t.} \quad f = Du \tag{1.12}$$

where

$$D = E_0(D) + \sum_{1\leqslant i\leqslant n} E_i(D)e_i + \sum_{1\leqslant i<j\leqslant n} E_{ij}(D)e_{ij} + \cdots + E_{1\cdots n}(D)e_{1\cdots n} \in (\mathbb{G}_n)^{N\times M} \tag{1.13}$$

$$u = E_0(u) + \sum_{1\leqslant i\leqslant n} E_i(u)e_i + \sum_{1\leqslant i<j\leqslant n} E_{ij}(u)e_{ij} + \cdots + E_{1\cdots n}(u)e_{1\cdots n} \in (\mathbb{G}_n)^{M} \tag{1.14}$$

$$f = E_0(f) + \sum_{1\leqslant i\leqslant n} E_i(f)e_i + \sum_{1\leqslant i<j\leqslant n} E_{ij}(f)e_{ij} + \cdots + E_{1\cdots n}(f)e_{1\cdots n} \in (\mathbb{G}_n)^{N} \tag{1.15}$$

Then, the l_1-norm minimization optimization problem in Eq.(1.12) can be rewritten as follows

$$\min k \in \mathbb{R}^+, \quad \text{s.t.} \quad f = Du, \quad \|u\|_1 \leqslant k \tag{1.16}$$

where $k = \sum_{l=1}^{M} k_l, k_l \in \mathbb{R}^+$, and $\mathbb{R}^+$ denotes the real number field which is positive. And the second constraint $\|u\|_1 \leqslant k$ can be transferred as

$$\begin{aligned}\|u\|_1 &= \sum_{1\leqslant l\leqslant M} |u_l| \\ &= \sum_{1\leqslant l\leqslant M} \sqrt{E_0^2(u_l) + \sum_{1\leqslant i\leqslant n} E_i^2(u_l) + \sum_{1\leqslant i<j\leqslant n} E_{ij}^2(u_l) + \cdots + E_{1\cdots n}^2(u_l)}\end{aligned}$$

$$= \sum_{1\leqslant l\leqslant M} \left\| \begin{matrix} E_0^2(u_l) \\ \vdots \\ E_i^2(u_l) \\ \vdots \\ E_{ij}^2(u_l) \\ \vdots \\ E_{1\cdots n}^2(u_l) \end{matrix} \right\|_2 \leqslant 1^{\mathrm{T}}k = 1^{\mathrm{T}}[k_1, k_2, \cdots, k_M]^{\mathrm{T}} = k \tag{1.17}$$

We substitute Eq.(1.17) into Eq.(1.16) to obtain

$$\forall l \in \{1, 2, \cdots, M\}, \quad \min_k 1^{\mathrm{T}}k \in \mathbb{R}, \quad \text{s.t.} \quad f = Du, \quad \left\| \begin{bmatrix} E_0^2(u_l) \\ \vdots \\ E_i^2(u_l) \\ \vdots \\ E_{ij}^2(u_l) \\ \vdots \\ E_{1\cdots n}^2(u_l) \end{bmatrix} \right\|_2 \leqslant k_l \tag{1.18}$$

Since

$$\begin{aligned}
f = Du \Leftrightarrow\ & 0 + \sum_{1\leqslant i\leqslant n} E_i(f)e_i + \sum_{1\leqslant i<j\leqslant n} E_{ij}(f)e_{ij} + \cdots + E_{1\cdots n}(f)e_{1\cdots n} \\
= & \left(E_0(D) + \sum_{1\leqslant i\leqslant n} E_i(D)e_i + \sum_{1\leqslant i<j\leqslant n} E_{ij}(D)e_{ij} + \cdots + E_{1\cdots n}(D)e_{1\cdots n}\right) \\
& \left(E_0(u) + \sum_{1\leqslant i\leqslant n} E_i(u)e_i + \sum_{1\leqslant i<j\leqslant n} E_{ij}(u)e_{ij} + \cdots + E_{1\cdots n}(u)e_{1\cdots n}\right) \\
= & \left(E_0(D)E_0(u) + \sum_{1\leqslant i\leqslant n} E_i(D)E_i(u) - \sum_{1\leqslant i<j\leqslant n} E_{ij}(D)E_{ij}(u)\right. \\
& \left. + \cdots + (-1)^{n(n-1)/2}E_{1\cdots n}(D)E_{1\cdots n}(u)\right)e_0 \\
& + \cdots + \left(\begin{matrix} E_t(D)E_0(u) + \dfrac{i-t}{|i-t|}\displaystyle\sum_{1\leqslant i\leqslant n} E_{ti}(D)E_i(u) \\ +E_0(D)E_t(u) + \dfrac{t-i}{|t-i|}\displaystyle\sum_{1\leqslant i\leqslant n} E_i(D)E_{it}(u) \\ + \cdots + (-1)^{(t-1)+(n(n-1)/2)}E_{1\cdots(t-1)(t+1)\cdots n}(D)E_{1\cdots n}(u) \end{matrix}\right)e_t
\end{aligned}$$

$$
+\cdots+\left(\begin{array}{c}
E_{ts}(D)E_0(u)+\dfrac{(i-t)(i-s)}{|i-t|\,|i-s|}\displaystyle\sum_{1\leqslant i\leqslant n}E_{tsi}(D)E_i(u)+E_t(D)E_s(u)\\
-E_s(D)E_t(u)+\dfrac{(i-t)(i-s)(j-t)(j-s)}{|i-t|\,|i-s|\,|j-s|\,|j-t|}\displaystyle\sum_{1\leqslant i<j\leqslant n}E_{tsij}(D)E_{js}(u)\\
+\dfrac{(j-t)(s-j)}{|j-t|\,|s-j|}\displaystyle\sum_{1\leqslant j\leqslant n}E_{tj}(D)E_{js}(u)+\cdots+(-1)^{(t-1)+(s-2)+(n(n-1)/2)}\\
E_{1\cdots(t-1)(t+1)\cdots(s-1)(s+1)\cdots n}(D)E_{1\cdots n}(u)
\end{array}\right)e_{ts}
$$

$$
+\cdots+\left(\begin{array}{c}
E_{1\cdots n}(D)E_0(u)+(-1)^{(n-i)}\displaystyle\sum_{1\leqslant i\leqslant n}E_{1\cdots(i-1)(i+1)\cdots n}(D)E_i(u)\\
+(-1)^{(2n-i-j+1)}\displaystyle\sum_{1\leqslant i<j\leqslant n}E_{1\cdots(i-1)(i+1)\cdots(j-1)(j+1)\cdots n}\\
(D)E_{ij}(u)+E_0(D)E_{1\cdots n}(u)
\end{array}\right)e_{1\cdots n}
\tag{1.19}
$$

where $e_0=1; t=1,\cdots,n; i\neq t, i\neq s, j\neq t, j\neq s; j>i, s>t; E_{ij}(D)=E_{ji}(D), E_{ij}(u)=E_{ji}(u)$. Therefore, we can obtain

$$
\forall l\in\{1,2,\cdots,M\},\quad \min_{\hat{u}}\hat{a}^{\mathrm{T}}\hat{u}\in\mathbb{R},\quad \text{s.t.}\quad \hat{f}=\hat{D}\hat{u},\quad \left\|\begin{bmatrix} E_0^2(u_l)\\ \vdots\\ E_i^2(u_l)\\ \vdots\\ E_{ij}^2(u_l)\\ \vdots\\ E_{1\cdots n}^2(u_l)\end{bmatrix}\right\|_2\leqslant k_l
\tag{1.20}
$$

$$
\hat{a}=[1,0,0,0,0,0,\cdots,1,0,0,0,0,0,\cdots]^{\mathrm{T}}\in\mathbb{R}^{(2^n+1)M}
\tag{1.21}
$$

$$
\begin{aligned}
\hat{u}=&[k_1,E_0(u_1),\cdots,E_i(u_1),\cdots,E_{ij}(u_1),\cdots,E_{1\cdots n}(u_1),\cdots,\\
&k_M,E_0(u_M),\cdots,E_i(u_M),\cdots,E_{ij}(u_M),\cdots,E_{1\cdots n}(u_M)]^{\mathrm{T}}\\
&\in\mathbb{R}^{(2^n+1)M},\quad 1\leqslant i<j\leqslant n
\end{aligned}
\tag{1.22}
$$

$$
\hat{f}=[E_0(f),\cdots,E_i(f),\cdots,E_{ij}(f),\cdots,E_{1\cdots n}(f)]^{\mathrm{T}}\in\mathbb{R}^{(2^n)N},\quad 1\leqslant i<j\leqslant n
\tag{1.23}
$$

$$
\hat{D}^{\mathrm{T}} = \begin{bmatrix}
0 & \cdots & 0 & \cdots & 0 & \cdots & 0 \\
\vdots & & \vdots & & \vdots & & \vdots \\
E_0(d_1) & \cdots & E_t(d_1) & \cdots & E_{ts}(d_1) & \cdots & E_{1\cdots n}(d_1) \\
\vdots & & \vdots & & \vdots & & \vdots \\
E_t(d_1) & \cdots & E_0(d_1) & \cdots & -E_s(d_1) & \cdots & (-1)^{(n-t)}E_{1\cdots(t-1)(t+1)n}(d_1) \\
\vdots & & \vdots & & \vdots & & \vdots \\
-E_{ts}(d_1) & \cdots & -E_s(d_1) & \cdots & E_0(d_1) & \cdots & (-1)^{(2n-t-s+1)}E_x(d_1) \\
\vdots & & \vdots & & \vdots & & \vdots \\
(-1)^{n(n-1)/2}E_{1\cdots n}(d_1) & \cdots & (-1)^{(t-1)+n(n-1)/2}E_z(d_1) & \cdots & yE_x(d_1) & \cdots & E_0(d_1) \\
\vdots & & \vdots & & \vdots & & \vdots \\
\vdots & & \vdots & & \vdots & & \vdots \\
0 & \cdots & 0 & \cdots & 0 & \cdots & 0 \\
\vdots & & \vdots & & \vdots & & \vdots \\
E_0(d_M) & \cdots & E_t(d_M) & \cdots & E_{ts}(d_M) & \cdots & E_{1\cdots n}(d_M) \\
\vdots & & \vdots & & \vdots & & \vdots \\
E_t(d_M) & \cdots & E_0(d_M) & \cdots & -E_s(d_M) & \cdots & (-1)^{(n-t)}E_{1\cdots(t-1)(t+1)n}(d_M) \\
\vdots & & \vdots & & \vdots & & \vdots \\
-E_{ts}(d_M) & \cdots & -E_s(d_M) & \cdots & E_0(d_M) & \cdots & (-1)^{(2n-t-s+1)}E_x(d_M) \\
\vdots & & \vdots & & \vdots & & \vdots \\
(-1)^{n(n-1)/2}E_{1\cdots n}(d_M) & \cdots & (-1)^{(t-1)+n(n-1)/2}E_z(d_M) & \cdots & yE_x(d_M) & \cdots & E_0(d_M)
\end{bmatrix} \tag{1.24}
$$

where $d_1, d_2, \cdots, d_M$ are the columns of measurement matrix D, $x = 1 \cdots (t-1)(t+1) \cdots (s-1)(s+1) \cdots n$; $y = (-1)^{(t-1)+(s-2)+n(n-1)/2}$; $z = 1 \cdots (t-1)(t+1) \cdots n$; $i \neq t, t \neq s, j \neq t, j \neq s$; $j > i, s > t$; $\hat{D} \in \mathbb{R}^{2^n N \times (2^n+1)M}$.

1.3.2 Noise Case

Now we focus on the l_1-norm minimization problem with noise, which is similar to the noiseless case. The minimization problem

$$\min \|u\|_1, \quad \text{s.t.} \quad \|Du - f\|_2 \leqslant \varepsilon \tag{1.25}$$

where $D \in (\mathbb{G}_n)^{N \times M}$, $u \in (\mathbb{G}_n)^M$ and $f \in (\mathbb{G}_n)^N$, can be reformulated as

$$\forall l \in \{1, 2, \cdots, M\}, \quad \min_{\hat{a}} \hat{a}^{\mathrm{T}} \hat{u} \in \mathbb{R}, \quad \text{s.t.} \quad \left\|\hat{D}\hat{u} - \hat{f}\right\|_2 \leqslant \varepsilon, \quad \left\| \begin{bmatrix} E_0^2(u_l) \\ \vdots \\ E_i^2(u_l) \\ \vdots \\ E_{ij}^2(u_l) \\ \vdots \\ E_{1\cdots n}^2(u_l) \end{bmatrix} \right\|_2 \leqslant k_l \tag{1.26}$$

Then

$$\left\|\hat{D}\hat{u} - \hat{f}\right\|_2 \leqslant \varepsilon \Leftrightarrow \hat{u}^{\mathrm{T}} \hat{D}^{\mathrm{T}} \hat{D} \hat{u} - 2\hat{u}^{\mathrm{T}} \hat{D}^{\mathrm{T}} \hat{f} + \hat{f}^{\mathrm{T}} \hat{f} \leqslant \varepsilon^2 \tag{1.27}$$

Or

$$\left\|\hat{D}\hat{u} - \hat{f}\right\|_2 \leqslant \varepsilon \quad \Leftrightarrow \quad \|z\|_2 \leqslant 2z_0 z_1 \tag{1.28}$$

where $z = \hat{D}\hat{u}$, $z_0 = 1$, $z_1 = \dfrac{\varepsilon^2}{2} + \hat{u}\hat{D}^{\mathrm{T}}\hat{f} - \dfrac{\hat{f}^{\mathrm{T}}\hat{f}}{2}$.

Here $\{z, z_0, z_1\}$ belongs to a rotated second-order cone. Considering the linear relationship of $\hat{u}$, z, z_0 and z_1, the following linear constraint can be obtained

$$\tilde{D}\tilde{u} = \tilde{b} \tag{1.29}$$

where

$$\tilde{u} = [\hat{u}, z, z_0, z_1] \in \mathbb{R}^{2^n N + (2^n+1)M + 2} \tag{1.30}$$

$$\tilde{b} = \left[0, 1, \frac{\hat{f}^{\mathrm{T}}\hat{f} - \varepsilon^2}{2}\right] \in \mathbb{R}^{2^n N + 2} \tag{1.31}$$

$$\tilde{D}=\begin{bmatrix} \hat{f}^{\mathrm{T}}\hat{D} & -1 & & \\ 0 & & \ddots & \\ \hat{D} & & & -1 \end{bmatrix} \in \mathbb{R}^{(2^n N+2)\times(2^n N+(2^n+1)M+2)} \tag{1.32}$$

Then we can convert (1.25) to a second-order cone problem as follows

$$\forall l \in \{1,2,\cdots,M\}, \quad \min_{\tilde{u}} \tilde{a}^{\mathrm{T}}\tilde{u} \in \mathbb{R}, \quad \text{s.t.} \quad \tilde{b}=\tilde{D}\tilde{u}, \quad \left\| \begin{bmatrix} E_0^2(u_l) \\ \vdots \\ E_i^2(u_l) \\ \vdots \\ E_{ij}^2(u_l) \\ \vdots \\ E_{1\cdots n}^2(u_l) \end{bmatrix} \right\|_2 \leqslant k_l,$$

$$\|z\|_2 \leqslant 2z_0z_1, \quad \tilde{c}=[\hat{c},0] \in \mathbb{R}^{2^n N+(2^n+1)M+2}, \quad \tilde{u}=[\hat{u},z,z_0,z_1] \in \mathbb{R}^{2^n N+(2^n+1)M+2} \tag{1.33}$$

It is clearly that Eq.(1.20) and Eq.(1.33) are both the standard forms of the SOCP problem and then can be solved by SeDuMi toolbox[29]. Therefore, the recovered 2^n-dimensional signal u' can be obtained by $\hat{u}$ or $\tilde{u}$ successfully.

1.4 Multi-dimensional Signal Processing in $\mathbb{G}_2$, $\mathbb{G}_3$ Space

1.4.1 Multi-dimensional Signal Processing in $\mathbb{G}_2$ Space

In $\mathbb{G}_2$ space, the l_1-norm minimization problem for 4-dimensional signals based on GA using SOCP can be formally described as follows.

Firstly, we define

$$D=E_0(D)+E_1(D)e_1+E_2(D)e_2+E_{12}(D)e_{12} \in (\mathbb{G}_2)^{N\times M} \tag{1.34}$$

$$u=E_0(u)+E_1(u)e_1+E_2(u)e_2+E_{12}(u)e_{12} \in (\mathbb{G}_2)^{M} \tag{1.35}$$

$$f=E_0(f)+E_1(f)e_1+E_2(f)e_2+E_{12}(f)e_{12} \in (\mathbb{G}_2)^{N} \tag{1.36}$$

where $E(D) \in \mathbb{R}$, $E(u) \in \mathbb{R}$ and $E(f) \in \mathbb{R}$.

Akin to Eq.(1.16) the following l_1-norm minimization problem

$$\min \|u\|_1, \quad \text{s.t.} \quad f=Du \tag{1.37}$$

can be rewritten as follows

$$\min k \in \mathbb{R}^+, \quad \text{s.t.} \quad f = Du, \|u\|_1 \leqslant k \tag{1.38}$$

Finally, we can obtain

$$\forall l \in \{1,2,\cdots,M\}, \quad \min_{\hat{u}} \hat{a}^{\mathrm{T}}\hat{u} \in \mathbb{R}, \quad \text{s.t.} \quad \hat{f} = \hat{D}\hat{u}, \quad \left\| \begin{bmatrix} E_0^2(u_l) \\ E_1^2(u_l) \\ E_2^2(u_l) \\ E_{12}^2(u_l) \end{bmatrix} \right\|_2 \leqslant k_l \tag{1.39}$$

where

$$\hat{a} = [1,0,0,0,0,\cdots,1,0,0,0,0]^{\mathrm{T}} \in \mathbb{R}^{5M} \tag{1.40}$$

$$\begin{aligned} \hat{u} = [&k_1, E_0(u_1), E_1(u_1), E_2(u_1), E_{12}(u_1), \cdots, \\ &k_M, E_0(u_M), E_1(u_M), E_2(u_M), E_{12}(u_M)]^{\mathrm{T}} \in \mathbb{R}^{5M} \end{aligned} \tag{1.41}$$

$$\hat{f} = [E_0(f), E_1(f), E_2(f), E_{12}(f)]^{\mathrm{T}} \in \mathbb{R}^{4N} \tag{1.42}$$

$$\hat{D} = \begin{bmatrix} 0 & 0 & 0 & 0 \\ E_0(d_1) & E_1(d_1) & E_2(d_1) & E_{12}(d_1) \\ E_1(d_1) & E_0(d_1) & -E_{12}(d_1) & -E_2(d_1) \\ E_2(d_1) & E_{12}(d_1) & E_0(d_1) & E_1(d_1) \\ -E_{12}(d_1) & -E_2(d_1) & E_1(d_1) & E_0(d_1) \\ \vdots & \vdots & \vdots & \vdots \\ 0 & 0 & 0 & 0 \\ E_0(d_M) & E_1(d_M) & E_2(d_M) & E_{12}(d_M) \\ E_1(d_M) & E_0(d_M) & -E_{12}(d_M) & -E_2(d_M) \\ E_2(d_M) & E_{12}(d_M) & E_0(d_M). & E_1(d_M) \\ -E_{12}(d_M) & -E_2(d_M) & E_1(d_M) & E_0(d_M) \end{bmatrix}^{\mathrm{T}} \in \mathbb{R}^{4N\times 5M} \tag{1.43}$$

1.4.2 Multi-dimensional Signal Processing in $\mathbb{G}_3$ Space

Similarly, in $\mathbb{G}_3$ space, the l_1-norm minimization problem for 8-dimensional signals can be obtained as follows

$$\forall l \in \{1,2,\cdots,M\}, \quad \min_{\hat{u}} \hat{a}^{\mathrm{T}}\hat{u} \in \mathbb{R}, \quad \text{s.t.} \quad \hat{f} = \hat{D}\hat{u}, \quad \left\| \begin{bmatrix} E_0^2(u_l) \\ E_1^2(u_l) \\ E_2^2(u_l) \\ E_3^2(u_l) \\ E_{12}^2(u_l) \\ E_{13}^2(u_l) \\ E_{23}^2(u_l) \\ E_{123}^2(u_l) \end{bmatrix} \right\|_2 \leqslant k_l \tag{1.44}$$

where

$$\hat{a} = [1,0,0,0,0,0,0,0,0,\cdots,1,0,0,0,0,0,0,0,0]^{\mathrm{T}} \in \mathbb{R}^{9M} \tag{1.45}$$

$$\begin{aligned}\hat{u} = [&k_1, E_0(u_1), E_1(u_1), E_2(u_1), E_3(u_1)E_{12}(u_1), E_{13}(u_1), E_{23}(u_1), E_{123}(u_1), \cdots, \\ &k_M, E_0(u_M), E_1(u_M), E_2(u_M), E_3(u_M)E_{12}(u_M), E_{13}(u_M), \\ &E_{23}(u_M), E_{123}(u_M)]^{\mathrm{T}} \in \mathbb{R}^{9M}\end{aligned} \tag{1.46}$$

$$\hat{f} = [E_0(f), E_1(f), E_3(f), E_{12}(f), E_{13}(f), E_{23}(f), E_{123}(f)]^{\mathrm{T}} \in \mathbb{R}^{8N} \tag{1.47}$$

$$\hat{D} = \left[\begin{array}{cccc}
0 & 0 & 0 & 0 \\
E_0(d_1) & E_1(d_1) & E_2(d_1) & E_3(d_1) \\
E_1(d_1) & E_0(d_1) & -E_{12}(d_1) & -E_{13}(d_1) \\
E_2(d_1) & E_{12}(d_1) & E_0(d_1) & -E_{23}(d_1) \\
E_3(d_1) & E_{13}(d_1) & E_{23}(d_1) & E_0(d_1) \\
-E_{12}(d_1) & -E_2(d_1) & E_1(d_1) & -E_{123}(d_1) \\
-E_{13}(d_1) & -E_3(d_1) & E_{123}(d_1) & E_1(d_1) \\
-E_{23}(d_1) & -E_{123}(d_1) & -E_3(d_1) & E_2(d_1) \\
-E_{123}(d_1) & -E_{23}(d_1) & E_{13}(d_1) & -E_{12}(d_1) \\
\vdots & \vdots & \vdots & \vdots \\
0 & 0 & 0 & 0 \\
0 & 0 & 0 & 0 \\
E_0(d_M) & E_1(d_M) & E_2(d_M) & E_3(d_M) \\
E_1(d_M) & E_0(d_M) & -E_{12}(d_M) & -E_{13}(d_M) \\
E_2(d_M) & E_{12}(d_M) & E_0(d_M) & -E_{23}(d_M) \\
E_3(d_M) & E_{13}(d_M) & E_{23}(d_M) & E_0(d_M) \\
-E_{12}(d_M) & -E_2(d_M) & E_1(d_M) & -E_{123}(d_M) \\
-E_{13}(d_M) & -E_3(d_M) & E_{123}(d_M) & E_1(d_M) \\
-E_{23}(d_M) & -E_{123}(d_M) & -E_3(d_M) & E_2(d_M) \\
-E_{123}(d_M) & -E_{23}(d_M) & E_{13}(d_M) & -E_{12}(d_M)
\end{array}\right.$$

$$\left.\begin{array}{cccc}
0 & 0 & 0 & 0 \\
E_{12}(d_1) & E_{13}(d_1) & E_{23}(d_1) & E_{123}(d_1) \\
-E_2(d_1) & -E_3(d_1) & E_{123}(d_1) & E_{23}(d_1) \\
E_1(d_1) & -E_{123}(d_1) & -E_3(d_1) & -E_{13}(d_1) \\
E_{123}(d_1) & E_1(d_1) & E_2(d_1) & E_{12}(d_1) \\
E_0(d_1) & -E_{23}(d_1) & E_{13}(d_1) & E_3(d_1) \\
E_{23}(d_1) & E_0(d_1) & -E_{12}(d_1) & -E_2(d_1) \\
-E_{13}(d_1) & E_{12}(d_1) & E_0(d_1) & E_1(d_1) \\
E_3(d_1) & -E_2(d_1) & E_1(d_1) & E_0(d_1) \\
\vdots & \vdots & \vdots & \vdots \\
0 & 0 & 0 & 0 \\
0 & 0 & 0 & 0 \\
E_{12}(d_M) & E_{13}(d_M) & E_{23}(d_M) & E_{123}(d_M) \\
-E_2(d_M) & -E_3(d_M) & E_{123}(d_M) & E_{23}(d_M) \\
E_1(d_M) & -E_{123}(d_M) & -E_3(d_M) & -E_{13}(d_M) \\
E_{123}(d_M) & E_1(d_M) & E_2(d_M) & E_{12}(d_M) \\
E_0(d_M) & -E_{23}(d_M) & E_{13}(d_M) & E_3(d_M) \\
E_{23}(d_M) & E_0(d_M) & -E_{12}(d_M) & -E_2(d_M) \\
-E_{13}(d_M) & E_{12}(d_M) & E_0(d_M) & E_1(d_M) \\
E_3(d_M) & -E_2(d_M) & E_1(d_M) & E_0(d_M)
\end{array}\right]^{\mathrm{T}}$$

$$\in \mathbb{R}^{8N \times 9M} \tag{1.48}$$

1.5 Experiments Results and Analysis

For validation of the proposed algorithm, we present numerical experiments that suggest empirical bounds on sparsity t relative to N (the number of measurements) for a multi-dimensional signal u to be perfectly recovered. The results can be seen as practical guidelines for scenarios where one can expect perfect recovery from random Gaussian GA multivector information using SOCP.

Our experiments aim at the recovery of the 4-dimensional signal and 8-dimensional signal. One shall note that our model in fact can be extended to even higher order, though we use the cases of 4 and 8-dimension as embodiment for empirical study.

1.5.1 4-dimensional Signal Reconstruction in $\mathbb{G}_2$ Space

1. Noiseless Case

Under a noiseless scene, numerical experiments in $\mathbb{G}_2$ space are conducted as follows:

1) Random Gaussian measurement multivector $D \in (\mathbb{G}_2)^{N\times M}(N \leqslant M)$.

Set the length of input 4-dimensional signal $M = 256$, N is the number of measurements, and then generate D with random entries sampled from independent and identically distributed (i.i.d.) Gaussian process with zero mean and variance equal to 1.

2) Sparse 4-dimensional input signal $u \in (\mathbb{G}_2)^M$.

Select a support set T of size $|T| = t$ (sparsity) uniformly at random, and sample a vector u on T with i.i.d. Gaussian entries.

3) 4-dimensional output signal f: multiplied D by input signal u to obtain f.

4) Construct $\hat{u}$, $\hat{f}$, and $\hat{D}$ (see from Eq.(1.39) to Eq.(1.42)).

5) To solve SOCP problem in Eq.(1.38), we use the SeDuMi toolbox[16]. And we also compute the error $\|u' - u\|_2$, where o is the recovered signal and u is the input signal.

6) We perform 100 rounds of experiments for each pair of sparsity t and the number of measurements N. These errors are saved and the number of perfect recovered experiments is also counted. When the error $\|u' - u\|_2 \leqslant 10^{-6}$, we think the signal is perfectly recovered.

The successful recovered results are shown in Fig.1.1, the components of original

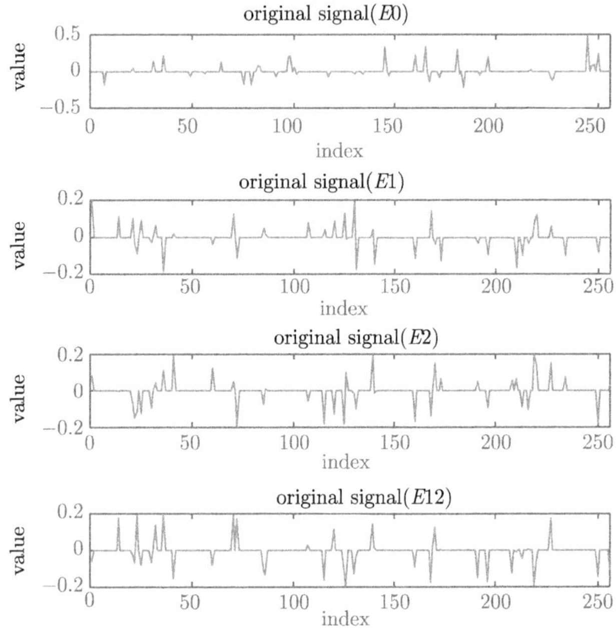

(a) Original 4-dimensional signal u with $t = 40$

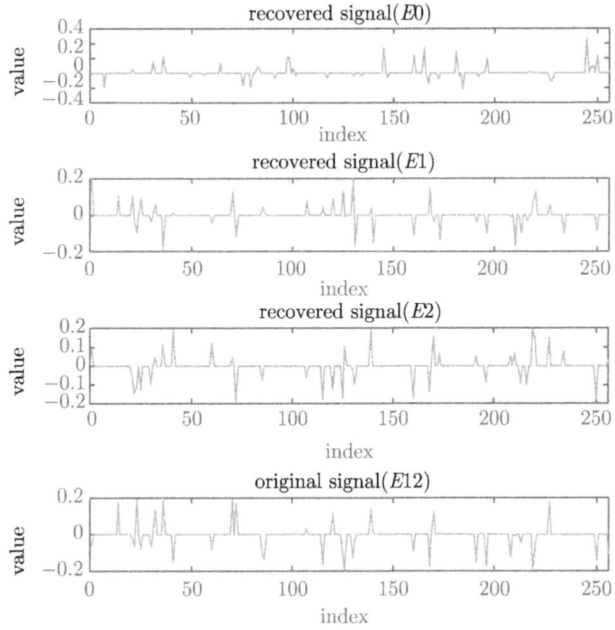

(b) The perfectly recovered signal u'

Fig.1.1 Recovery experiment for l_1-norm minimization for 4-dimensional signals ($M = 256$, $N = 128$)

4-dimensional signal $u \in (\mathbb{G}_2)^M$ of size $M = 256$ are listed on the left, and the components of recovered 4-dimensional signal u' are on the right. The number of measurements N is set as 128 and the sparsity $t = 40$.

To further illustrate the correlation between the percentage of successful recovery and sparsity level t for different number of measurements $N = 48, 96, 128, 196, 256$, we perform 800 rounds of experiments for 4-dimensional signals recovery, and the results are shown in Fig.1.2. We discover that, for a fixed sparsity level t, the recovery probability increases as we take more measurements. Moreover, for the 4-dimensional signals, under the same number of measurements N, with the increase number of sparsity level t, the perfect recovery probability decreases conversely. The larger measurements number is, the more obvious descending trend becomes.

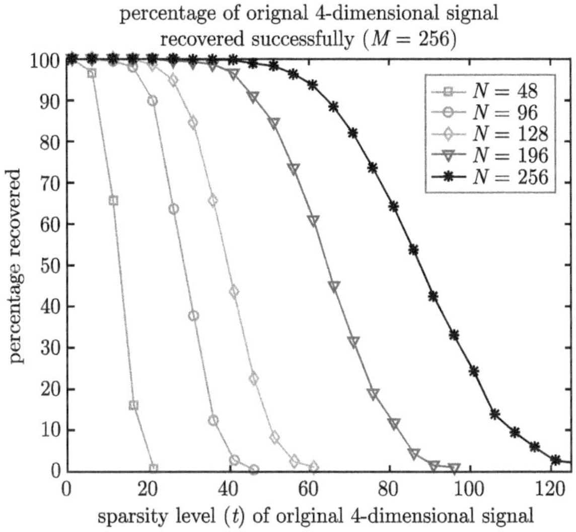

Fig.1.2 The rate of 800 4-dimensional signals correctly recovered as a function of the sparsity level t for different numbers N of measurements under $M = 256$

2. Noise Case

It is worthy to mention that the reconstruction experiment in $\mathbb{G}_2$ space with the noisy 4-dimensional input signal, can also be regarded as the denoising experiment. In this noise case, we set the length of input 4-dimensional signal $M = 256$, the number of measurements $N = 128$. The measurements are all corrupted by the additive zero-mean white Gaussian noise vector in $\mathbb{G}_2$ space, consisting of i.i.d. with standard deviation σ. Then the squared l_2-norm of the noise vector $\|e\|_2^2$ is a chisquare random variable with mean $4N\sigma^2$ and variance $8N\sigma^4$. Here, the constraint parameter ε is set to be $\sigma\sqrt{4N+4\sqrt{2N}}$.

To further illustrate the robustness of the proposed algorithm to noise, ten rounds of 4-dimensional signal recovery experiments at different noise levels are conducted and Table 1.1 shows the average recovery errors of 4-dimensional signal in $\mathbb{G}_2$ space with regard to corresponding noise level at different sparsity.

The results in Table 1.1 illustrate that the proposed recovery algorithm is robust to noise contamination. From Table 1.1, we can conclude that with the increase of the disturbed additive Gaussian noise, the recovery performance tends to be more efficient under the same sparsity level. Moreover, at the same noise level, the recovery results are better with the increasing sparsity level.

Table 1.1 Average 4-dimensional signal recovery errors with different noise level σ at different sparsity level t

σ \ t	2	4	8	16	32	64	100
0.01	1.9×10^{-5}	3.9×10^{-5}	4.6×10^{-5}	1.1×10^{-4}	0.02	1.66	2.51
0.03	9.6×10^{-5}	2.2×10^{-4}	7.5×10^{-4}	8.7×10^{-4}	0.02	1.76	2.61
0.05	5.9×10^{-4}	7.7×10^{-4}	9.6×10^{-4}	1.1×10^{-3}	0.02	1.72	3.26
0.1	1.1×10^{-3}	3.6×10^{-3}	4.8×10^{-3}	0.02	0.07	1.98	2.89
0.3	0.01	0.04	0.04	0.14	0.20	1.61	2.70
0.5	0.14	0.16	0.35	0.42	0.74	1.60	2.61

1.5.2 8-dimensional Signal Reconstruction in $\mathbb{G}_3$ Space

1. Noiseless Case

Similar to the 4-dimensional signal reconstruction experiments in noiseless scenario shown above, the recovery of 8-dimensional signal can be described as follows:

1) Random Gaussian measurement multivector $D\in(\mathbb{G}_3)^{N\times M}(N\leqslant M)$.

2) Sparse 8-dimensional input signal $u\in(\mathbb{G}_3)^M$.

3) 8-dimensional output signal f: multiplied D by input signal u to obtain f.

4) Construct $\hat{u}$, $\hat{f}$, and $\hat{D}$ (see from Eq.(1.44) to Eq.(1.47)).

5) Use the SeDuMi toolbox to solve SOCP problem in Eq.(1.43) and compute the error.

6) We also perform 100 rounds of experiments for each pair of sparsity t and the number of measurements N. These errors are saved and the number of perfect recovery experiments is also counted. We claim that the reconstruction is perfect if the error $\|u' - u\|_2 \leqslant 10^{-6}$.

The results are shown in Fig.1.3, where the components of original 8-dimensional signal $u \in (\mathbb{G}_3)^M$ of size $M = 256$ are listed on the left, and the components of

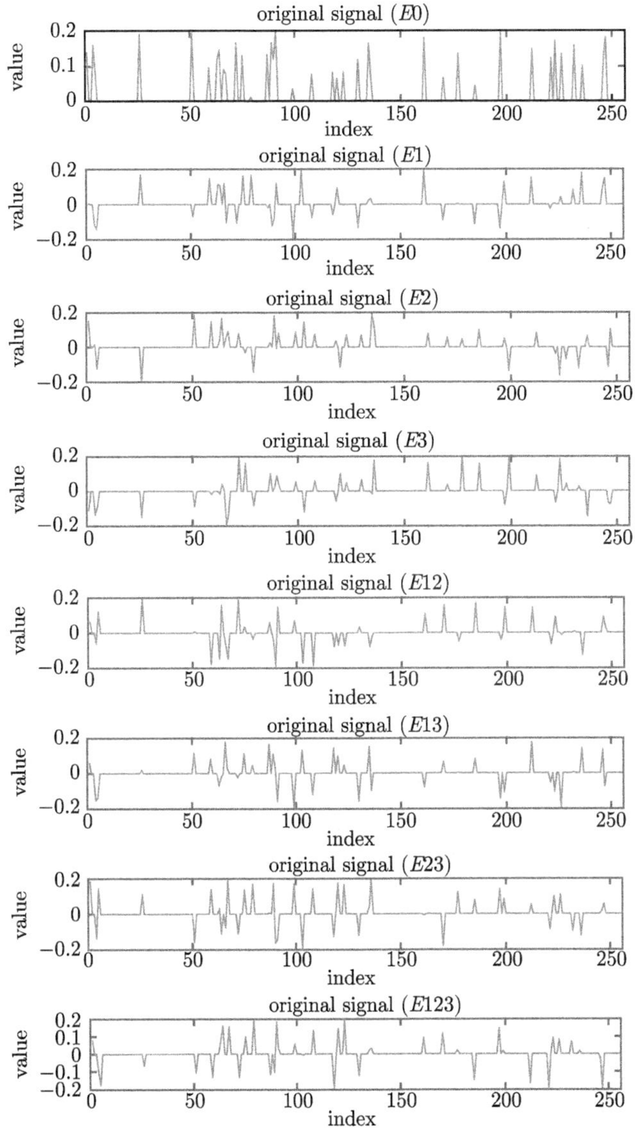

(a) Original 8-dimensional signal u with $t = 40$

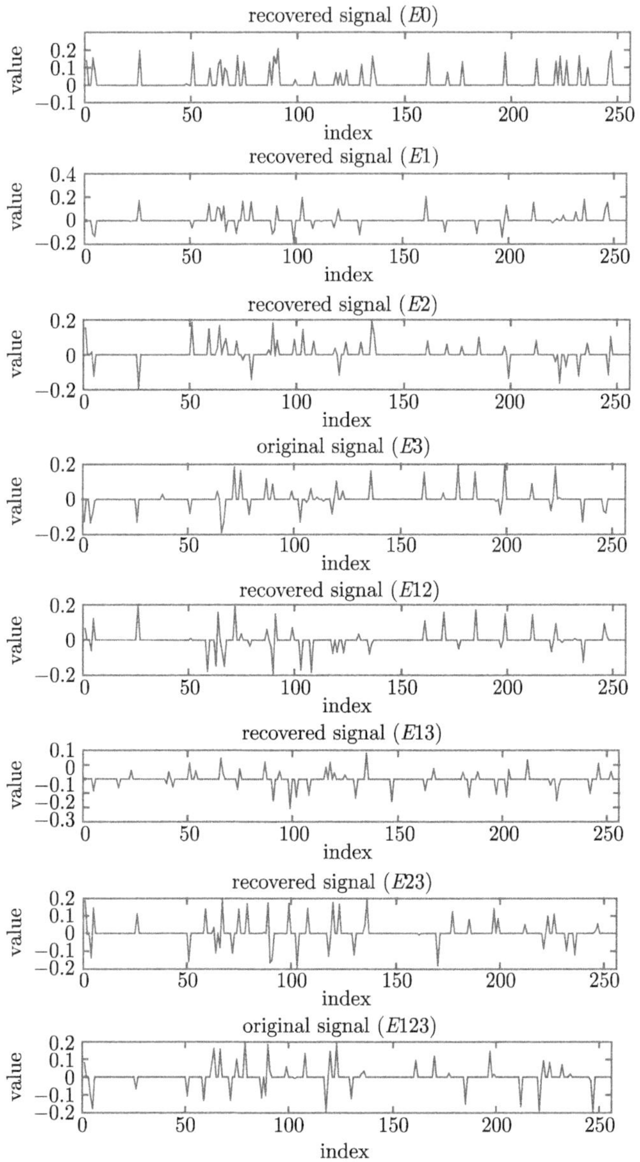

(b) The perfectly recovered signal u'

Fig.1.3 Recovery experiment for l_1-norm minimization for 8-dimensional signals ($M = 256,\ N = 128$)

recovered 8-dimensional signal $\hat{u}$ are on the right. The number of measurements N is set as 128 and the sparsity $t = 40$.

Fig.1.4 shows the relationship between the percentage successful recovered and the number of measurements N under different sparsity level $t = 4, 12, 20, 28, 36$. We perform 100 8-dimensional signals recovery experiments, and the results are shown in Fig.1.4. Each curve represents a different sparsity level t. As expected, when the number of nonzero components increases, more measurements are necessary to guarantee signal recovery.

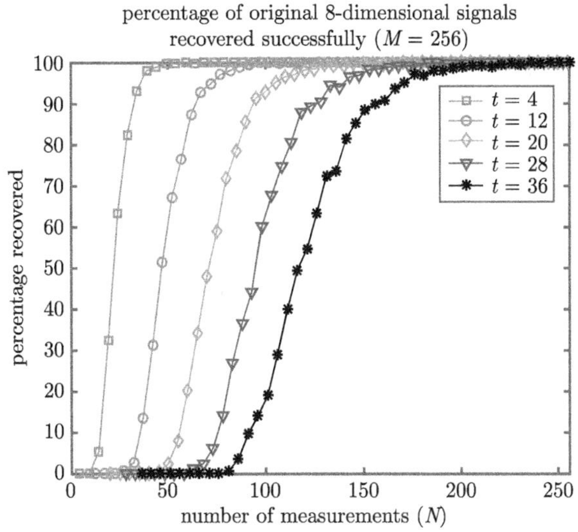

Fig.1.4 The rate of 800 8-dimensional signals correctly recovered as a function of the sparsity level t for different numbers N of measurements under $M = 256$

2. Noise Case

Similar to the noise case of input 4-dimensional signal in $\mathbb{G}_2$ space, the length of 8-dimensional signal M, the number of measurements N are set to be 256 and 128, respectively. The measurements are corrupted by the additive zero-mean white Gaussian noise vector in $\mathbb{G}_3$ space, comprised of i.i.d. with standard deviation σ. Then the squared l_2-norm of the noise vector $\|e\|_2^2$ is a chisquare random variable with mean $8N\sigma^2$ and variance $16N\sigma^4$.

With the measurements in $\mathbb{G}_3$ space disturbed by additive Gaussian noise, we carry out ten rounds of 8-dimensional signal recovery experiments at different noise levels to further show the robustness of the proposed algorithm to noise. Table 1.2 shows the average recovery errors of 8-dimensional signal in $\mathbb{G}_3$ space with regard to corresponding noise level at different sparsity. As shown in Table 1.2, the recovery errors are getting smaller with the decreasing noise levels and sparsity level.

Table 1.2 Average 8-dimensional signal recovery errors with different noise level σ at different sparsity level t

σ \ t	2	4	8	16	32	64	100
0.01	3.2×10^{-5}	5.4×10^{-5}	9.8×10^{-5}	1.1×10^{-4}	0.03	1.77	3.64
0.03	2.4×10^{-4}	2.9×10^{-4}	4.7×10^{-4}	7.0×10^{-4}	0.05	2.09	3.71
0.05	9.4×10^{-4}	1.4×10^{-3}	1.7×10^{-3}	4.2×10^{-3}	0.06	2.10	3.84
0.1	2.3×10^{-3}	3.7×10^{-3}	6.4×10^{-3}	0.02	0.19	2.14	3.85
0.3	0.03	0.07	0.08	0.10	0.44	2.44	3.89
0.5	0.22	0.42	0.61	0.63	1.57	2.69	4.33

To further instinctively analyze the effect of noise and sparsity on 8-dimensional signal recovery errors, averaged recovery errors of 8-dimensional signal as a function of noise and sparsity are plotted in Fig.1.5 for different values of noise level σ and sparsity level t.

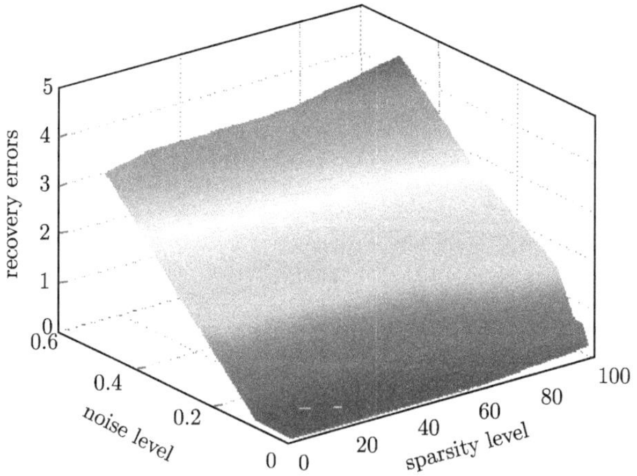

Fig.1.5 Averaged recovery errors of 8-dimensional signals as a function of noise level σ and sparsity level t

The tendency of recovery errors with respect to noise and sparsity are clearly highlighted in Fig.1.5. The recovery errors are small with both small values of σ and t, and the results are greater for larger parameters. In summary, the proposed algorithm can be seen as a good method that is able to perfectly and stably recovering multi-dimensional signals from random Gaussian measurements.

1.6 Conclusions

In this chapter, we propose an algorithm using geometric algebra (GA) for solving the l_1-norm minimization problem of multi-dimensional signals, which can be converted to SOCP and then solved by SeDuMi toolbox. The algorithm represents the multi-dimensional signal as a multivector in the GA form. Numerical examples of 4-dimensional signals and 8-dimensional signals are provided to illustrate the feasibility and effectiveness of the algorithm. The main feature of the proposed algorithm is that the optimization of GA multivector value is converted to the optimization of real values, whereby many mature toolboxes can be applied. The l_1-norm minimization is successfully extended to high dimensional signals, and can find potential applications in multi-dimensional signal processing and analysis. Since conformal geometric algebra[21] is regarded as an integration of advanced geometric invariants and covariant systems based on conformal model and geometric algebra, at next steps, we will focus on multi-dimensional signal processing using conformal geometric algebra.

References

[1] Candès E J, Romberg J, Tao T. Robust uncertainty principles: Exact signal reconstruction from highly incomplete frequency information. IEEE Trans. Inf. Theory, 2006, 52: 489-509.

[2] Donoho L. Compressed sensing. IEEE Trans. Inf. Theory, 2006, 52: 1289-1306.

[3] Candès E J, Romberg J, Tao T. Stable signal recovery from incomplete and inaccurate measurements. Commun. Pure Appl. Math., 2006, 59: 1207-1223.

[4] Winter S, Kellermann W, Sawada H, et al. MAP-based underdetermined blind source separation of convolutive mixtures by hierarchical clustering and l_1-norm minimization. EURASIP J. Adv. Signal Process, 2007: 1-12.

[5] Ender J H G. On compressive sensing applied to radar. Signal Process, 2010, 90(5): 1402-1414.

[6] Yu S, Khwaja A S, Ma J. Compressed sensing of complex-valued data. Signal Process, 2012, 92: 357-362.

[7] Hamilton W R. On quaternions. Proc. Roy. Irish Acad., 1844: 1-16.

[8] Ell T, Sangwine S. Hypercomplex Fourier transforms of color images. IEEE Trans. Image Process, 2007, 16: 22-35.

[9] Vía J, Palomar D P, Vielva L. Generalized likelihood ratios for testing the properness of quaternion Gaussian vectors. IEEE Trans. Signal Process, 2011, 59: 1356-1370.

[10] Vía J, Ramírez D, Vielva L. Properness and widely linear processing of quaternion random vectors. IEEE Trans. Inf. Theory, 2010, 56(7): 3502-3515.

[11] Jiang T, Wei M. Equality constrained least squares problem over quaternion field. Appl. Math. Lett, 2003, 16: 883-888.

[12] Vía J, Palomar D P, Vielva L, et al. Quaternion ICA from second-order statistics. IEEE Trans. Signal Process, 2011, 59(4): 1586-1600.

[13] Javidi S, Took C C, Mandic D P. A fast independent component analysis algorithm for quaternion signals. IEEE Trans. Neural Networks, 2010, 22(12): 1967-1978.

[14] Bayro Corrochano E. Theory and use of the quaternion wavelet transform. Journal of Mathematical Imaging and Vision, 2006, 24: 19-35.

[15] Xu Z, Song W J, Wang X Q, et al. L_1-norm minimization for quaternion signals. Journal of Southeast University, 2013, 29(1): 33-37.

[16] Sturm J F. Using SeDuMi 1.02, a MATLAB toolbox for optimization over symmetric cones. Optim. Meth. Softw., 1998, 11-12: 625-653.

[17] Franchini S, Gentile A, Sorbello F, et al. Conformal ALU: A conformal geometric algebra coprocessor for medical image processing. IEEE Transactions on Computers, 2015, 64(4): 955-970.

[18] Su H, Bo Z. Conformal geometric algebra based band selection and classification for hyperspectral imagery. 2016 8th Workshop on Hyperspectral Image and Signal Processing: Evolution in Remote Sensing (WHISPERS), Los Angeles, CA, 2016: 1-4.

[19] Pham M T, Yoshikawa T, Furuhashi T, et al. Robust feature extractions from geometric data using geometric algebra. IEEE International Conference on Systems Man and Cybernetics, 2009: 529-533.

[20] Gunn C. A new approach to euclidean plane geometry based on projective geometric algebra. Mathematics, 2015, 704(4): 41-58.

[21] Hitzer E, Nitta T, Kuroe Y. Applications of Clifford's geometric algebra. Adv. Appl. Clifford Algebras, 2013, 23(2): 377-404.

[22] Lopez Gonzalez G, Altamirano Gomez G, Bayro Corrochano E. Geometric entities voting schemes in the conformal geometric algebra framework. Adv. Aplic. Clifford Algebra, 2016, 26(3): 1045-1059.

[23] Zhou S L, Valchev D G, Dinovitser A, et al. Dispersion-independent terahertz classification based on geometric algebra for substance detection. 2016 41st International Conference on Infrared, Millimeter, and Terahertz Waves (IRMMW-THz), Copenhagen, 2016: 1, 2.

[24] Gunn G. A new approach to euclidean plane geometry based on projective geometric algebra. Mathematics, 2015, 704(4): 41-58.

[25] Wareham R, Cameron J, Lasenby J. Applications of conformal geometric algebra in computer vision and graphics. LNCS 3519, 2004: 329-349.

[26] Lopes W B, Al-Nuaimi A, Lopes C G. Geometric-algebra LMS adaptive filter and its application to rotation estimation. IEEE Signal Processing Letters, 2016, 23(6): 858-862.

[27] Ebling J, Scheuermann G. Clifford Fourier transform on vector fields. IEEE Transactions on Visualization & Computer Graphics, 2005, 11(4): 469-479.

[28] Wang R, Zhou Y X, Jin Y L, et al. Sparse fast Clifford Fourier transform. Frontiers of Information Technology & Electronic Engineering, 2017, 18(8): 1131-1141.

[29] Li Y, Liu W, Li X, et al. GA-SIFT: A new scale invariant feature transform for multispectral image using geometric algebra. Information Sciences, 2014: 559-572.

[30] Shen M M, Wang R, Cao W M. Joint sparse representation model for multi-channel image based on reduced geometric algebra. IEEE Access. Image Process, 2018, 6: 24213-24223.

[31] Bayro Corrochano E, Zang Y. The motor extended Kalman filter a geometric approach for rigid motion estimation. Journal of Mathematical Imaging and Vision, 2000, 13: 205-228.

[32] Wang R, Xiang G J, Zhang F. L_1-norm minimization for octonion signals. International Conference on Audio, Language and Image Processing. IEEE, 2017: 552-556.

Chapter 2

GA-SVD: A Novel Singular Value Decomposition Algorithm for Multispectral Image Based on Geometric Algebra

2.1 Introduction

In recent years, multispectral images, combined with spectral and spatial information, reflect more optical characteristics of the scene than traditional chromaticity diagrams, and have shown a wide range of applications in the fields of aerospace, remote sensing, environmental monitoring, biomedicine and military surveillance[1−3]. Specifically, a multispectral image is commonly composed of numerous images captured by sensors over a variety of discrete bands, and it records the response values of a specific sensor under a specific light. Since the spectral and spatial information contained in a multispectral image can be acquired simultaneously, multispectral images have ability to take into account both spatial image and spectral analysis. The provided abundant spectral and spatial information can be applied in not only the qualitative and quantitative analysis of observed targets, but also the physical structure detection of the target[4,5]. Consequently, the multispectral image analysis has attracted much attention and become hot topics lately[6−8].

Currently, the singular value decomposition (SVD), which has been one of the most efficient and powerful tools in numerical algebra and has wide applications in image processing tasks[9,10]. It decomposes the signal into a series of superposition of linear components with good stability and invariance. Since singular values and the corresponding vectors can reflect the intrinsic properties of the original data and preserve the relationship of rows or columns in the matrix, respectively, the SVD shows its great effectiveness and potential in many image applications, such as linear discriminant analysis, sparse representation and principal component analysis[11−13].

Recently, the SVD of quaternion matrix has achieved the state-of-the-art performance in terms of color image analysis. Zhang[14] performed the SVD of quaternion matrix theoretically and demonstrated the existence of the quaternion SVD by describing an isomorphism between a quaternion matrix and its equivalent complex matrix. A novel method for vector-sensor signal modelling and processing has been put forward by Bihan et al. taking advantage of quaternion algebra theory[15]. Jie et al.[16] provided a color image energy model in which the optimal number of singular values for reconstruction of the noisy color image can be obtained by calculating the curve of the color image energy. However, there are very few works focusing on the SVD of multichannel signals, which are presented as multispectral images typically.

Fortunately, geometric algebra (GA), gives a powerful computing framework in signal and image analysis tasks and with the formidable way it provides for multichannel signals, a great number of researches has been explored in multi-dimensional space[17−19]. Ebling et al.[20] extended the Fourier transform in (multi) vector fields using GA and further provided the generalizations of correlation, convolution, and derivative theorems. Wang et al.[21] presented a novel algorithm namely sparse fast Clifford Fourier transform (SFCFT) to make great advancement in computing performance over scalar and vector fields. Li et al.[22] proposed an efficient SIFT algorithmic framework based on GA for multispectral images to avoid losing a great deal of spectral information and distinctive feature points.

Since the current leading SVD algorithms are mainly exploited for gray-scale or color images, they cannot work efficiently for multispectral images. We focus in this chapter on the presentation of a novel SVD algorithm using geometric algebra (GA), namely GA-SVD, for multispectral images by converting it to the SVD of its real representation. With the theory of GA, a GA multivector can be used to represent a multispectral image to successfully make follow-up processing in a holistic manner and completely retain inner-relationship of different spectral channels on multispectral image. Combined with the formidable GA theory and the classical

SVD framework, the GA-SVD can be performed to preserve both spectral and spatial information.

The remainder of this chapter is organized as follows. Section 2.2 recalls some preliminaries about the geometric algebra and reviews the process of real SVD methods. Section 2.3 describes the proposed GA-SVD algorithm for multispectral image. Section 2.4 gives examples in $\mathbb{G}_2$ and $\mathbb{G}_3$ space. Compression and denoising experiments for multispectral image are provided in Section 2.5. Section 2.6 concludes the chapter.

2.2 Related Work

2.2.1 The Basics of Geometric Algebra

In this section, we briefly review the basics of GA, which was introduced by William K. Clifford and also called Clifford algebra[16–25]. Suppose $\mathbb{G}_n$ is n-dimensional GA with an orthonormal basis of vectors, which leads to a basis

$$\{1, \{e_i\}, \{e_i e_j\}, \cdots, \{e_1 e_2 \cdots e_n\}\} \tag{2.1}$$

Specifically, $\mathbb{G}_n$ can be represented by $\mathbb{G}_{p,q}$, where $n = p + q$ and p, q are the number of vectors with positive square and negative square in the basis of the space, respectively. That is

$$e_i^2 = \begin{cases} 1, & 1 \leqslant i \leqslant p \\ -1, & p+1 \leqslant i \leqslant n \end{cases} \tag{2.2}$$

In this chapter, we focus on $\mathbb{G}_n$, in which $q = 0$. The geometric product of two basis is anti-commutative, and

$$e_i e_j = e_{ij} = -e_j e_i = -e_{ji}, \quad i, j = 1, \cdots, n, \quad i \neq j \tag{2.3}$$

$$e_i^2 = 1, \quad i = 1, \cdots, n \tag{2.4}$$

$$e_i e_{ij} = e_i e_i e_j = e_j, \quad i, j = 1, \cdots, n, \quad i \neq j \tag{2.5}$$

For vectors ω and z of $\mathbb{G}_n$, the geometric product is defined as:

$$wz = w \cdot z + w \wedge z \tag{2.6}$$

where $w \cdot z$ denotes the inner product, $w \wedge z$ denotes the outer product. Since the e_1 vectors are orthogonal, then $e_i e_j = e_i \cdot e_j + e_i \wedge e_j = e_i \wedge e_j$.

Take $\mathbb{G}_3$ for instance, the orthogonal bases are constructed by vectors with $2^3 = 8$ grades, which is given by: $\{1, \{e_1, e_2, e_3\}, \{e_1e_2, e_1e_3, e_2e_3\}, \{e_1e_2e_3\}\}$ and a simpler form: $\{1, e_1, e_2, e_3, e_{12}, e_{13}, e_{23}, e_{123}\}$. For a $2n$-dimensional signal, only n basis is needed. Generally, the outer product of k vectors is called a k-blade. The number k is called the grade of the blade. The complete mathematical correlation between two vectors has been provided by GA. As the extension of vectors to higher dimensions, multivectors are the basic components in GA. Any multivector $M \in \mathbb{G}_n$ can be described by

$$M = E_0 + \sum_{1 \leqslant i \leqslant n} E_i(M)e_i + \sum_{1 \leqslant i < j \leqslant n} E_{ij}(M)e_{ij} + \cdots + E_{1\cdots n}(M)e_{1\cdots n} \tag{2.7}$$

where $E(M) \in \mathbb{R}$.

2.2.2 Singular Value Decomposition (SVD)

The singular value decomposition (SVD) was presented for the first time by Stewart et al.[26,27]. Now it becomes one of the most efficient and powerful tools in finding the eigenvalues and is widely employed in various image processing tasks, such as image compression and denoising. For any real matrix $A \in \mathbb{R}^{m \times n}$, there exist orthogonal matrices $U = [u_1, u_2, \cdots, u_m]V^{\mathrm{T}} = [v_1^{\mathrm{T}}, v_2^{\mathrm{T}}, \cdots, v_n^{\mathrm{T}}]$ and

$$\Sigma = \begin{bmatrix} \Sigma' & 0 \\ 0 & 0 \end{bmatrix} \tag{2.8}$$

where $\Sigma' = \mathrm{diag}\{\sigma_1, \sigma_2, \cdots, \sigma_p\} \in \mathbb{R}^{p \times p}$ and then

$$A = U\Sigma V^{\mathrm{T}} = \sum_{i=1}^{p} \sigma_i u_i v_i^{\mathrm{T}} \tag{2.9}$$

where σ_i are the positive singular values of A, $\sigma_1 \geqslant \sigma_2 \geqslant \cdots \geqslant \sigma_p > 0$ and $p = \min\{m, n\}$. More singular values will be required for the restoration of A, if A is an image of high frequency and worse noise.

2.3 The GA-SVD Algorithm for Multispectral Image

In this section, the process of GA-SVD algorithm for multispectral image is presented in detail.

2.3.1 Representation of Multispectral Image in GA

Each pixel of a multispectral image I can be denoted in $\mathbb{G}_n$ as

$$I(x,y) = 0 + \sum_{1\leqslant i\leqslant n} I_i(x,y)e_i + \sum_{1\leqslant i<j\leqslant n} I_{ij}(x,y)e_{ij} + \cdots + I_{1\cdots n}(x,y)e_{1\cdots n} \quad (2.10)$$

where $I_i(x,y), I_j(x,y), I_{1\cdots n}(x,y)$ are the spectral components of multispectral image pixel $I(x,y)$ at position (x,y) respectively.

Specifically, all the spectral channels of the multispectral image are assigned to the basis $\{1, \{e_i\}, \{e_ie_j\}, \cdots, \{e_1e_2\cdots e_n\}\}$ of the 2^n-dimensional GA ($\mathbb{G}_n$) respectively, and a GA multivector is utilized to present each multispectral image pixel $I(x,y)$ in GA.

Therefore, a multispectral image $F \in (\mathbb{G}_n)^{M\times N}$ with M rows and N columns can be represented as

$$F = 0 + \sum_{1\leqslant i\leqslant n} E_i(F)e_i + \sum_{1\leqslant i<j\leqslant n} E_{ij}(F)e_{ij} + \cdots + E_{1\cdots n}(F)e_{1\cdots n}, E(F) \in \mathbb{R}^{M\times N} \quad (2.11)$$

Then, the representation of multispectral image in GA has exactly attracted great attention currently. It has been widely employed in the field of multispectral image processing owing to its capability to handle the spectral channels in a holistic way and preserve the inner-relationship of the spectral channels successfully. Since our work is not the first one to represent multispectral images using GA, the representation method based on GA introduced in this chapter has some differences from the paper by Li et al.[22]. In the paper by Li et al., it only uses 1-blade to represent a multispectral image (see Eq.(14) in Ref.[22]), it says "There are two multivector in the result of the Eq.(27), and the second multivector cannot be used in the image processing simply." In this chapter, we reserve all the outer product parts, which brings the relationship of components in GA space into correspondence with the correlation of spectral channels in a multispectral image, and can be considered as a more comprehensive representation of a multispectral image. It can be applied in studying the problems of multispectral images processing due to its capability to treat the spectral channels holistically without losing the relationship of the spectral channels.

2.3.2 The Implementation of GA-SVD Algorithm

For a GA multivector $F \in (\mathbb{G}_n)^{M\times N}$ of a multispectral image, then there exist two GA unitary matrix $U \in (\mathbb{G}_n)^{M\times M}$ and $V \in (\mathbb{G}_n)^{N\times N}$ such that

$$U^H F V = \begin{bmatrix} \Sigma' & 0 \\ 0 & 0 \end{bmatrix} \equiv \Sigma \in \mathbb{R}^{M\times N} \tag{2.12}$$

where $\Sigma' = \mathrm{diag}\{\sigma_1, \sigma_2, \cdots, \sigma_p\} \in R^{p\times p}$, σ_i are the positive singular values of $\sigma_1 \geqslant \sigma_2 \geqslant \cdots \geqslant \sigma_p$ and $P = \min\{M, N\}$ and H represents the conjugate transpose operation.

That is, for any $M \times N$ GA multivector F, its SVD is obtained by

$$F = U\Sigma V^H = U \begin{bmatrix} \Sigma' & 0 \\ 0 & 0 \end{bmatrix} V^H \tag{2.13}$$

where $UU^H = I^{M\times M}$, $VV^H = I^{N\times N}$, I is an identity matrix.

Clearly, the GA-SVD of a GA multivector can be performed through different kinds of algorithms.

Suppose

$$F = \Bigg(E_0(F) + \sum_{1\leqslant i\leqslant n} E_i(F)e_i + \sum_{1\leqslant i<j\leqslant n} E_{ij}(F)e_{ij} + \cdots + E_{1\cdots n}(F)e_{1\cdots n} \in (G_n)^{M\times N} \Bigg)$$

where $E(F) \in \mathbb{R}^{M\times N}$, its real representation F_R can be denoted as follows

$$F_R = \begin{bmatrix} E_0(F) & \cdots & E_t(F) & \cdots & -E_{ts}(F) & \cdots & (-1)^{n(n-1)/2}E_{1\cdots n}(F) \\ \vdots & & \vdots & & \vdots & & \vdots \\ E_t(F) & \cdots & E_0(F) & \cdots & -E_s(F) & \cdots & (-1)^{(t-1)+(n(n-1)/2)}E_{1\cdots n}(F) \\ \vdots & & \vdots & & \vdots & & \vdots \\ E_{ts}(F) & \cdots & -E_s(F) & \cdots & E_0(F) & \cdots & (-1)^{(t-1)+(s-2)+(n(n-1)/2)}E_x(F) \\ \vdots & & \vdots & & \vdots & & \vdots \\ E_{1\cdots n}(F) & \cdots & (-1)^{(n-t)}E_y(F) & \cdots & (-1)^{(2n-t-s+1)}E_x(F) & \cdots & E_0(F) \end{bmatrix} \tag{2.14}$$

where $x = 1\cdots(t-1)(t+1)\cdots(s-1)(s+1)\cdots n; y = 1\cdots(t-1)(t+1)\cdots(s-1)(s+1)\cdots n; t = 1\cdots n; i \neq t, t \neq s, j \neq t, j \neq s; j > i, s > t; E_{ij} = E_{ji}$ and $F_R \in \mathbb{R}^{2^n M\times 2^n N}$.

The properties of F_R are shown as follows.

Theorem 1 Let F, $K \in (\mathbb{G}_n)^{M\times N}$, $A \in (\mathbb{G}_n)^{N\times S}$ and $c \in \mathbb{R}$. Then

1) $(F+K)_R = F_R + K_R, (cF)_R = cF_R, (FA)_R = F_R A_R$.

2) $(F^H)_R = (F_R)^{\mathrm{T}}$.

3) If $F \in (\mathbb{G}_n)^{M\times M}$ is a unitary matrix, F_R is an orthogonal matrix.

Furthermore, the correlation between the GA-SVD of a GA multivector and the SVD of its real representation are given as follows.

Theorem 2 Let $F \in (\mathbb{G}_n)^{M\times N}$, then the singular values of F_R repeat 2^n times.

Thus, if the SVD of a $M\times N$ GA multivector F is to be computed, we can perform the SVD of its $2^n M \times 2^n N$ real representation F_R as an alternative. Substantially, due to the orthogonal transformations, we can depict the standard form for the real representation of a GA multivector.

Theorem 3 Suppose that $F \in (\mathbb{G}_n)^{M\times N}$ and $F_R \in R^{2^n M\times 2^n N}$ is the real representation of F. Then there exist orthogonal matrices $U_R \in R^{2^n M\times 2^n M}$ and $V_R \in R^{2^n N\times 2^n N}$ such that

$$U_R^{\mathrm{T}} F_R V_R = \begin{bmatrix} \Sigma_R' & 0 \\ 0 & 0 \end{bmatrix} \equiv \Sigma_R \in \mathbb{R}^{2^n M\times 2^n N} \tag{2.15}$$

where $\Sigma_R' = \mathrm{diag}\{(\sigma_R)_1, (\sigma_R)_2, \cdots, (\sigma_R)_{P_R}\} \in R^{p_R\times p_R}$,$(\sigma_R)_i$ are the positive singular values of F, $(\sigma_R)_1 \geqslant (\sigma_R)_2 \geqslant \cdots \geqslant (\sigma_R)_{P_R}$, $P_R = \min\{2^n M, 2^n N\}$ and T represents the transpose operation. That is, for any $2^n M \times 2^n N$ real matrix F_R, its SVD is obtained by

$$F_R = U_R \Sigma_R V_R^{\mathrm{T}} = U_R \begin{bmatrix} \Sigma_R' & 0 \\ 0 & 0 \end{bmatrix} V_R^{\mathrm{T}} \tag{2.16}$$

where $UU^{\mathrm{T}} = I^{2^n M\times 2^n M}, VV^{\mathrm{T}} = I^{2^n N\times 2^n N}$, I is an identity matrix.

We now consider the relationship of F and its real representation F_R.

Theorem 4 Suppose that $F \in (\mathbb{G}_n)^{M\times N}$ and its real representation $F_R \in R^{2^n M\times 2^n N}$, we have $F = U\Sigma' V^H$, $F_R = U_R \Sigma_R' V_R^{\mathrm{T}}$ and for $1 \leqslant M \leqslant N, \Sigma' = \mathrm{diag}\{\sigma_1, \sigma_2, \cdots, \sigma_M\}$, for $1 \leqslant 2^n M \leqslant 2^n N, \Sigma_R' = \mathrm{diag}\{(\sigma_R)_1, (\sigma_R)_2, \cdots, (\sigma_R)_M\}$. Then

1) $\Sigma' = \mathrm{row}_{\mathrm{odd}}^{2^{n-1}}(\mathrm{col}_{\mathrm{odd}}^{2^{n-1}}(\Sigma'_R))$.

$$2)\ U = \left(E_0(U) + \sum_{1\leqslant i\leqslant n} E_i(U)e_i + \sum_{1\leqslant i<j\leqslant n} E_{ij}(U)e_{ij} + \cdots + E_{1\cdots n}(U)e_{1\cdots n}\right),$$

$U_R = [(U_R)_0 \cdots (U_R)_i \cdots (U_R)_{ij} \cdots (U_R)_{1\cdots n}]^{\mathrm{T}}$, then

$$U = \left((U_R)_0 - \sum_{1\leqslant i\leqslant n} (U_R)_i e_i - \sum_{1\leqslant i<j\leqslant n} (U_R)_{ij} e_{ij} - \cdots - (U_R)_{1\cdots n} e_{1\cdots n}\right).$$

$$3)\ V = \left(E_0(V) + \sum_{1\leqslant i\leqslant n} E_i(V)e_i + \sum_{1\leqslant i<j\leqslant n} E_{ij}(V)e_{ij} + \cdots + E_{1\cdots n}(V)e_{1\cdots n}\right),$$

$V_R = [(V_R)_0 \cdots (V_R)_i \cdots (V_R)_{ij} \cdots (V_R)_{1\cdots n}]^{\mathrm{T}}$, then

$$V = \left((V_R)_0 - \sum_{1\leqslant i\leqslant n} (V_R)_i e_i - \sum_{1\leqslant i<j\leqslant n} (V_R)_{ij} e_{ij} - \cdots - (V_R)_{1\cdots n} e_{1\cdots n}\right).$$

where $1 < i < j \leqslant n, \mathrm{row}_{\mathrm{odd}}^{2^{n-1}}(P), \mathrm{col}_{\mathrm{odd}}^{2^{n-1}}(P)$ extracts the odd rows and odd columns of matrix respectively for 2^{n-1} times.

2.3.3 The Reconstruction of Multispectral Image Based on GA-SVD

Actually, the GA-SVD of a GA multivector is obtained by the SVD of its real representation. Thus, the original multispectral image is reconstructed through the sum of the outer products:

$$F' = U\Sigma V^H = \sum_{i=1}^{M} \sigma_i u_i v_i^H \tag{2.17}$$

where F' is reconstruction of $F \in (\mathbb{G}_n)^{M\times N}$; $1 \leqslant M \leqslant N$, u_i and v_i are the column vectors of U and V, respectively, σ_i is the diagonal terms of real matrix Σ, U; Σ; V can be obtained by Theorem 4.

Usually, the singular values with large magnitude can present the most useful information of a multispectral image while the detailed information of the multispectral image can be provided by the remaining smaller ones. Then we can provide a more universal reconstruction form as

$$F'_K = U\Sigma_K V^H = \sum_{i=1}^{K} \sigma_i u_i v_i^H \tag{2.18}$$

where Σ_K is obtained by remaining the largest K singular values and set the next $M-K$ to 0, F'_K is reconstruction of F with K singular values, U; Σ_K; V can be obtained by Theorem 4.

2.4 The SVD Algorithm in $\mathbb{G}_2$, $\mathbb{G}_3$ Space

In this section, the process of GA-SVD algorithm in $\mathbb{G}_2$, $\mathbb{G}_3$ space is presented in detail.

2.4.1 The SVD Algorithm in $\mathbb{G}_2$ Space

The proposed GA-SVD algorithm in $\mathbb{G}_2$ space can be then obtained, and can be better adapted to color images. According to Section 2.1, a three channels color image B' with M rows and N columns can be represented as a 2-dimensional GA multivector, which is shown as follows:

$$B' = 0 + E_1(B)e_1 + E_2(B)e_2 + E_{12}(B)e_{12} \tag{2.19}$$

where $E_1(B), E_2(B), E_{12}(B) \in \mathbb{R}^{M\times N}$. Therefore, its SVD can be obtained as follows

$$B = U\Sigma V^H = U\begin{bmatrix} \Sigma' & 0 \\ 0 & 0 \end{bmatrix} V^H \tag{2.20}$$

where $\Sigma' = \operatorname{diag}\{\sigma_1, \sigma_2, \cdots, \sigma_p\} \in \mathbb{R}^{p\times p}$, V^H represent the conjugate transpose of V, $U \in (\mathbb{G}_2)^{M\times M}, V \in (\mathbb{G}_2)^{N\times N}$.

And its real representation can be denoted as follows[27–29],

$$B_R \equiv \begin{bmatrix} 0 & -E_1(B) & -E_2(B) & -E_{12}(B) \\ E_1(B) & 0 & -E_{12}(B) & E_2(B) \\ E_2(B) & E_{12}(B) & 0 & -E_1(B) \\ E_{12}(B) & -E_2(B) & E_1(B) & 0 \end{bmatrix} \tag{2.21}$$

where the singular values of B_R appear in fours. That is to say, the SVD of a $M\times N$ 2-dimensional GA multivector where the singular values of B_R appear in fours. That is to say, the SVD of a $M\times N$ 2-dimensional GA multivector B can be computed by the results obtained by SVD of the $4M\times 4N$ real representation matrix B_R. For the real matrix B_R, according to Eq.(2.9), there exist orthogonal matrices $U_R \in \mathbb{R}^{4M\times 4M}$ and $V_R \in \mathbb{R}^{4N\times 4N}$ such that

$$B_R = U_R\Sigma_R V_R^{\mathrm{T}} = U_R\begin{bmatrix} \Sigma'_R & 0 \\ 0 & 0 \end{bmatrix} V_R^{\mathrm{T}} \tag{2.22}$$

The relation is defined as follows:

1) $\Sigma' = \text{row}^2_{\text{odd}}(\text{col}^2_{\text{odd}}(\Sigma'_R))$.

2) $U = E_0(U) + E_1(U)e_1 + E_2(U)e_2 + E_{12}(U)e_{12}$,

$U_R = \begin{bmatrix} (U_R)_0 & (U_R)_1 & (U_R)_2 & (U_R)_{12} \end{bmatrix}^{\mathrm{T}}$, then

$V = (V_R)_0 - (U_V)_1 e_1 - (V_R)_2 e_2 - (U_V)_{12} e_{12}$.

3) $V = E_0(V) + E_1(V)e_1 + E_2(V)e_2 + E_{12}(V)e_{12}$.

$U_V = [(U_V)_0 \quad (U_V)_1 \quad (U_V)_2 \quad (U_V)_{12}]^{\mathrm{T}}$, then

$V = (V_R)_0 - (U_V)_1 e_1 - (V_R)_2 e_2 - (U_V)_{12} e_{12}$

where $\text{row}^2_{\text{odd}}(P)$, $\text{col}^2_{\text{odd}}(P)$ extracts the odd rows and odd columns of matrix P respectively for twice.

2.4.2 The SVD Algorithm in $\mathbb{G}_3$ Space

Similarly, inspired by performing the GA-SVD of the 2-dimensional GA multivector, the framework for computing the GA-SVD of the 3-dimensional GA multivector can be also achieved using its equivalent real matrices.

For a 3-dimensional GA multivector $A \in (\mathbb{G}_3)^{M\times N}, A = E_0(A) + E_1(A)e_1 + E_2(A)e_2 + E_3(A)e_3 + E_{12}(A)e_{12} + E_{13}(A)e_{13} + E_{23}(A)e_{23} + E_{123}(A)e_{123}$, there exist two orthogonal matrices $U \in (\mathbb{G}_3)^{M\times M}, V \in (\mathbb{G}_3)^{N\times N}$. Such that

$$U^H A V = \begin{bmatrix} \Sigma' & 0 \\ 0 & 0 \end{bmatrix} \equiv \Sigma \in \mathbb{R}^{M\times N} \tag{2.23}$$

where $\Sigma' = \text{diag}\{\sigma_1, \sigma_2, \cdots, \sigma_r\} \in \mathbb{R}^{r\times r}$, σ_i are the positive singular values of A, $\sigma_1 \geqslant \sigma_2 \geqslant \cdots \geqslant \sigma_r$, and $r = \min\{m, n\}$. H represents the conjugate transpose operation.

That is, for any $M \times N$ 3-dimensional GA multivector $A \in (\mathbb{G}_3)^{M\times N}$, its SVD is obtained by

$$A = U\Sigma V^H = U \begin{bmatrix} \Sigma' & 0 \\ 0 & 0 \end{bmatrix} V^H \tag{2.24}$$

We can transform the GA-SVD of the 3-dimensional GA multivector to the SVD of its equivalent real matrix. For the 3-dimensional GA multivector $A \in (\mathbb{G}_3)^{M\times N}$, its equivalent real matrix $A_R \in \mathbb{R}^{8M\times 8N}$ can be obtained as follows:

$$A_R = \begin{bmatrix} E_0(A) & -E_1(A) & -E_2(A) & -E_3(A) & -E_4(A) & -E_5(A) & -E_6(A) & -E_7(A) \\ E_1(A) & E_0(A) & -E_3(A) & E_2(A) & -E_5(A) & E_4(A) & E_7(A) & -E_6(A) \\ E_2(A) & E_3(A) & E_0(A) & -E_1(A) & -E_6(A) & -E_7(A) & E_4(A) & E_5(A) \\ E_3(A) & -E_2(A) & E_1(A) & E_0(A) & -E_7(A) & E_6(A) & -E_5(A) & E_4(A) \\ E_4(A) & E_5(A) & E_6(A) & E_7(A) & E_0(A) & -E_1(A) & -E_2(A) & -E_3(A) \\ E_5(A) & -E_4(A) & E_7(A) & -E_6(A) & E_1(A) & E_0(A) & E_3(A) & -E_2(A) \\ E_6(A) & -E_7(A) & -E_4(A) & E_5(A) & E_2(A) & -E_3(A) & E_0(A) & E_1(A) \\ E_7(A) & E_6(A) & -E_5(A) & -E_4(A) & E_3(A) & E_2(A) & -E_1(A) & E_0(A) \end{bmatrix} \tag{2.25}$$

Thus, the SVD of a 3-dimensional GA multivector $A \in (\mathbb{G}_3)^{M\times N}$ can be performed by dealing with the SVD of its real representation matrix $A_R \in \mathbb{R}^{8M\times 8N}$.

For the real matrix A_R, according to Eq.(2.9), there exist orthogonal matrices $U_R \in \mathbb{R}^{8M\times 8M}$ and $V_R \in \mathbb{R}^{8N\times 8N}$ such that

$$A_R = U_R \varSigma_R V_R^{\mathrm{T}} = U_R \begin{bmatrix} \varSigma_R' & 0 \\ 0 & 0 \end{bmatrix} V_R^{\mathrm{T}} \tag{2.26}$$

The relation is defined as follows

1) $\varSigma' = \mathrm{row}_{\mathrm{odd}}^4(\mathrm{col}_{\mathrm{odd}}^4(\varSigma_R'))$.

2) $U = E_0(U) + E_1(U)e_1 + E_2(U)e_2 + E_3(U)e_3 + E_{12}(U)e_{12} + E_{13}(U)e_{13} + E_{23}(U)e_{23} + E_{123}(U)e_{123}$, $U_R = [(U_R)_0 \quad (U_R)_1 \quad (U_R)_2 \quad (U_R)_3 \quad (U_R)_{12} \quad (U_R)_{13} \quad (U_R)_{23} \quad (U_R)_{123}]^{\mathrm{T}}$, then $U = (U_R)_0 - (U_R)_1 e_1 - (U_R)_2 e_2 - (U_R)_3 e_3 - (U_R)_{12} e_{12} - (U_R)_{13} e_{13} - (U_R)_{23} e_{23} - (U_R)_{123} e_{123}$.

3) $V = E_0(V) + E_1(V)e_1 + E_2(V)e_2 + E_3(V)e_3 + E_{12}(V)e_{12} + E_{13}(V)e_{13} + E_{23}(V)e_{23} + E_{123}(V)e_{123}$, $V_R = [(V_R)_0 \quad (V_R)_1 \quad (V_R)_2 \quad (V_R)_3 \quad (V_R)_{12} \quad (V_R)_{13} \quad (V_R)_{23} \quad (V_R)_{123}]^{\mathrm{T}}$, then $V = (V_R)_0 - (V_R)_1 e_1 - (V_R)_2 e_2 - (V_R)_3 e_3 - (V_R)_{12} e_{12} - (V_R)_{13} e_{13} - (V_R)_{23} e_{23} - (V_R)_{123} e_{123}$.

Where $\mathrm{row}_{\mathrm{odd}}^4(P)$, $\mathrm{col}_{\mathrm{odd}}^4(P)$ extracts the odd rows and odd columns of matrix P respectively for four times.

2.5 Experimental Analysis

2.5.1 Data Sets

The Columbia Multispectral Image Database[30] is utilized in our experiments. This multispectral image data set contains 32 real-world scenes of a various real world materials and objects, each with spatial resolution 512×512 and spectral resolution 31, which includes full spectral resolution reflectance data captured from 400nm to 700nm in 10nm steps. More details can be found in http://www.cs.columbia.edu/CAVE/databases/multispectral/.

In this chapter, several arbitrary multispectral images are chosen to estimate the performance of the proposed GA-SVD method, and three of them are shown in Fig.2.1 as examples. We choose 6 bands randomly from each selected multispectral image to form 20 pseudo color images, respectively. Fig.2.1(a)-(r) show the images of band 2, 5, 12, 20, 25 and 28, respectively. Fig.2.1(s)-(x) show the pseudo color images in which band 2, 5, and 12 images and 20, 25, and 28 images are input

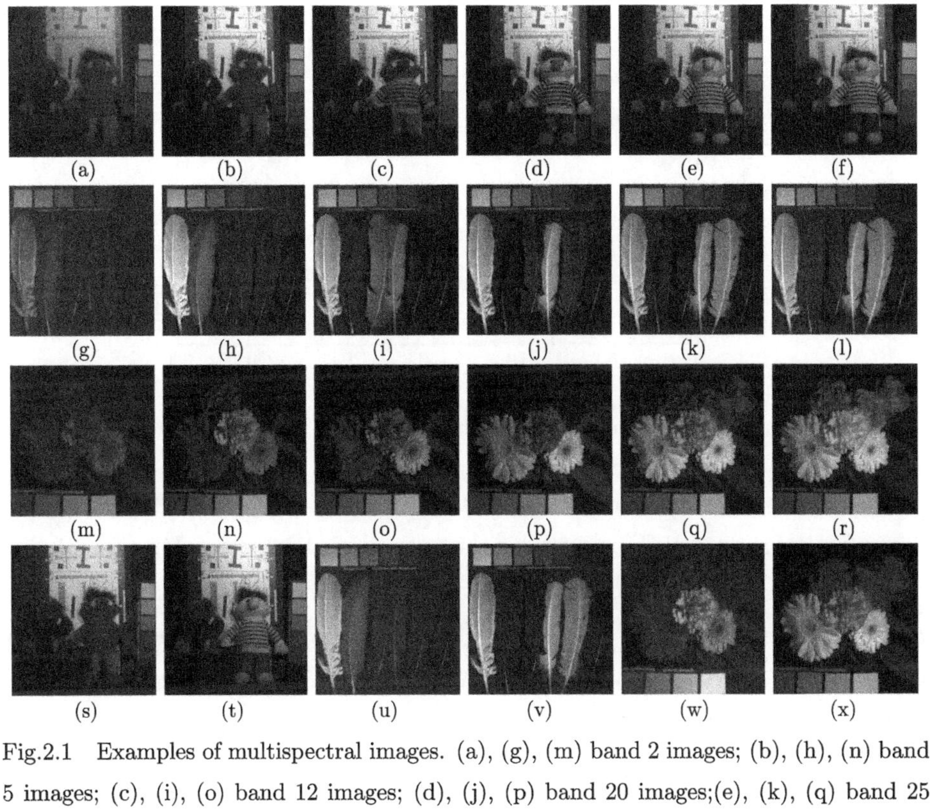

Fig.2.1 Examples of multispectral images. (a), (g), (m) band 2 images; (b), (h), (n) band 5 images; (c), (i), (o) band 12 images; (d), (j), (p) band 20 images;(e), (k), (q) band 25 images; (f), (l), (r) band 28 images; (s), (u), (w) the pseudo color images using 2, 5, and 12 band images as the RGB input; (t), (v), (x) the pseudo color images using 20, 25, and 28 band images as the RGB input

as the R, G and B channels, respectively. Then, they are utilized to conduct the multispectral image compression and denoising experiments.

2.5.2 Multispectral Image Compression

The compression experiments of multispectral image[30] are carried out in this section for validity evaluation of the proposed GA-SVD and conventional real SVD algorithms. The singular values distribution obtained by the proposed GA-SVD and conventional real SVD methods of multispectral images in Fig.2.1 are shown in Fig.2.2.

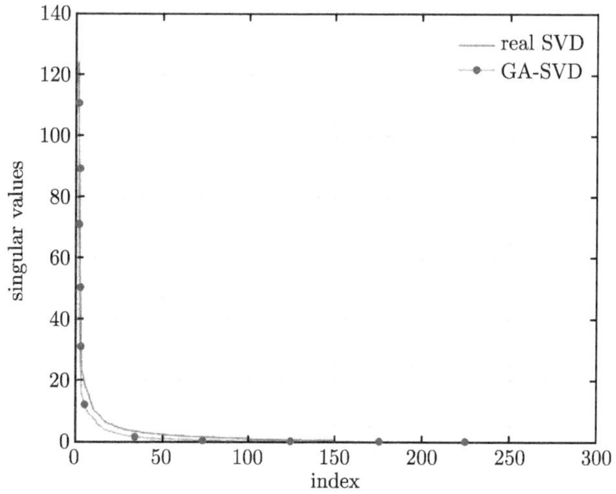

Fig.2.2 Singular values of the testing multispectral image

It is clearly that the singular values obtained by the proposed GA-SVD method decay further more rapid, compared to the conventional real case. Therefore, in line with Eq.(2.20), we can represent the multispectral image approximately through the sum of the preceding K singular values. That is to say, less number of singular values are necessary for the proposed GA-SVD method to reconstruct the original multispectral image, and more specifically in Fig.2.2, when retaining the largest 50 singular values approximately can basically preserve most of the image information while nearly the largest 150 singular values are needed for the real SVD method to achieve better reconstruction. Generally, the original multispectral image can be great estimated using just several small K. Multispectral image in Fig.2.1 is taken as the original image, shown in (a) and (b) of Fig.2.3. The compressed images with $K = 2, 4, 8, 16, 48, 100$ using the proposed GA-SVD method are shown in (c)-(h) and (o)-(t) of Fig.2.3 and accordingly, images shown in (i)-(n) and (u)-(z) of Fig.2.3 are compressed images based on real SVD algorithm.

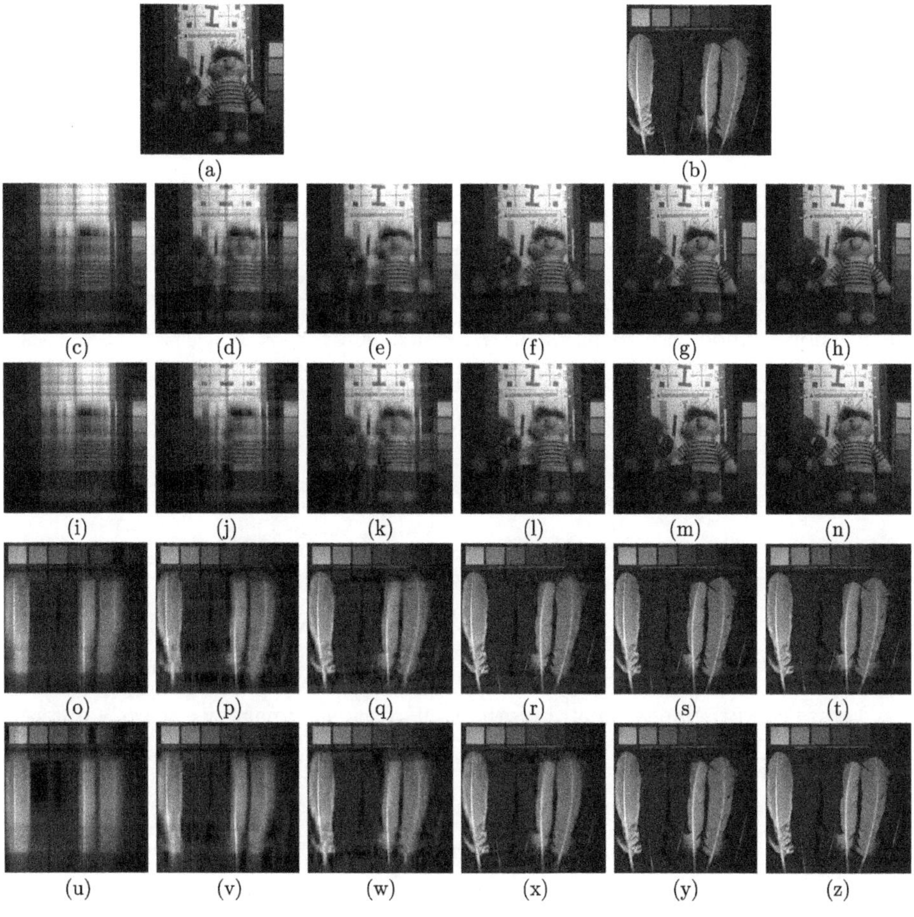

Fig.2.3 Original and compressed multispectral images. (a) and (b) original pseudo color images (band 20, 25, 28); (c)-(h) and (o)-(t) compressed images with $k = 2, 4, 8, 16, 48, 100$ using the proposed method; (i)-(n) and (u)-(z) compressed images with $k = 2, 4, 8, 16, 48, 100$ using the real SVD method

It can be clearly seen from Fig.2.3 that the compressed multispectral image using our GA-SVD algorithm has much better results while choosing relatively small number of K values. Specifically, more details and marginal information are preserved using our GA-SVD method. For explanation of the better performance achieved by the proposed GASVD method towards multispectral image compression experiments, we owe it to the utilization of GA theory. Specifically, taking advantage of GA theory, the multispectral image is efficiently represented as a GA multivector and then it is able to be processed in a holistic way, which avoids losing of the relationship of multiple channels in multispectral image. Moreover, once enough singular values are selected, the performance of the two methods seems to be equivalent, both very close to the original multispectral image.

2.5.3 Multispectral Image Denoising

Then, we conduct another application for multispectral image-denoising experiments at different noise levels over various multispectral images. Practically, under an additive noise model, we focus on a GA multivector $Z = F + W$, where F is a multispectral image, W is a random Gaussian noise in the GA form. Then the denoising result F^K of F is obtained using our GA-SVD algorithm as

$$F^K = U\Sigma^K V^H = U \begin{bmatrix} (\Sigma')^K & 0 \\ 0 & 0 \end{bmatrix} V^H \tag{2.27}$$

where $(\Sigma')^K = \mathrm{diag}\{\sigma_1, \sigma_2, \cdots, \sigma_K\}$is given by remaining the largest K singular values and set the next $M - K$ to $0, U$; Σ^K; V can be obtained by Theorem 4.

We first consider the influence of the number of singular values. Since large singular values can provide the main information of a noisy image, we select a different number of singular values in this experiment to evaluate the effect of the number of valid singular values. The multispectral image shown in Fig.2.1 is disturbed by additive Gaussian white noise with standard deviation 25 ($\delta = 25$). To demonstrate the effectiveness of our proposed algorithm, more visually, under the same condition, we employ the real SVD for each single spectral channel. Fig.2.4 illustrates the denoising results with different K values.

The reconstruction results become more obvious and significant with increasing number of singular values. However, if K is chosen to be very small, most noise would be filtered, and the original image would be lost accordingly. Thus, the most important step is the determination of the optimal K value to greatly eliminate noise. It is clearly shown in Fig.2.4 that the real SVD method, which represents each spectral channel independently as a monochrome image to use real SVD, loses some inherent structures and spectral information due to the loss of inner-relationship of spectral channels in multispectral image. Our GA-SVD method regards the total multi-spectral image as a GA multivector. It shows superiority especially in homogenous region, provides more particulars of the image, and present spectral information in a vivid manner after denoising.

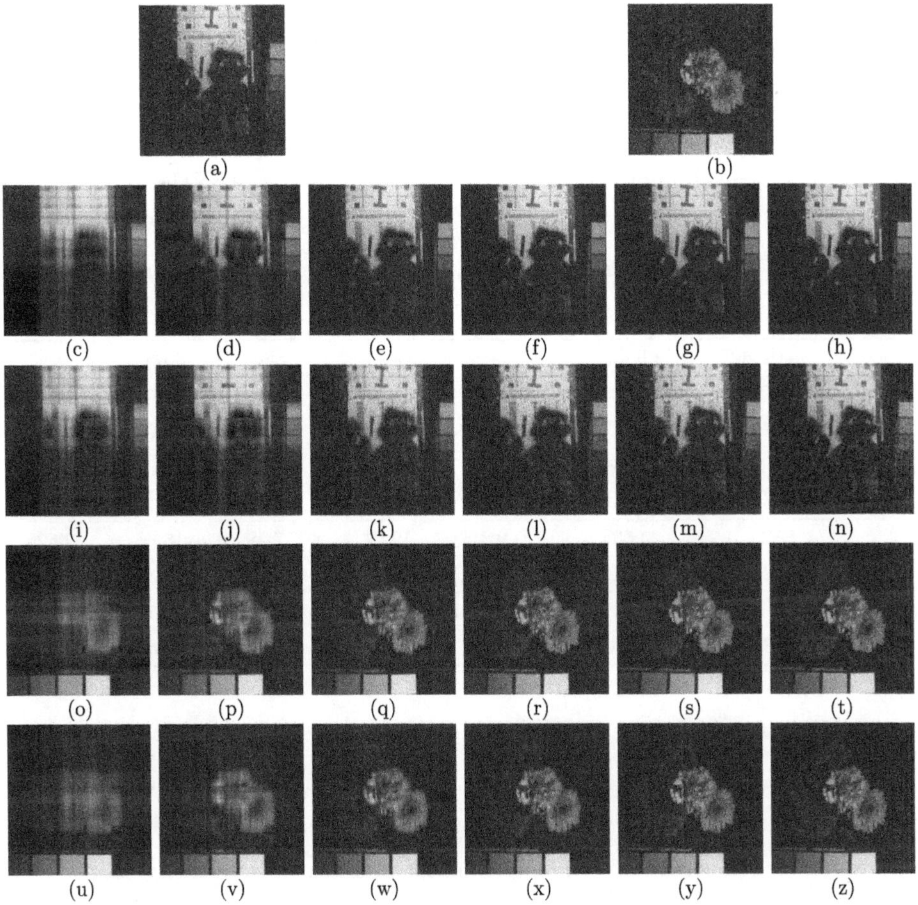

Fig.2.4 Denoising results. (a) and (b) original noisy pseudo color images (band 2, 5, 12); (c)-(h) and (o)-(t) denoising images with $k = 2, 5, 10, 20, 50, 100$ using the proposed method; (i)-(n) and (u)-(z) denoising images with $k = 2, 5, 10, 20, 50, 100$ using the real SVD method

The noisy multispectral images are obtained by adding white Gaussian noise with different standard deviation $\delta = 15, 25, 35, 45$. To further evaluate the higher performance of the proposed GA-SVD algorithm compared with the traditional real SVD method in an objective way, the PSNR values are given for example. These quantitative evaluation results further illustrate the conclusion in an objective way. The PSNR values using conventional real SVD and our GA-SVD methods for different testing multispectral images over different noise levels are listed in Table 2.1. As can be seen from Table 2.1, our GA-SVD treats the spectral channels of the multispectral image holistically without losing spectral information and leads to much better results. The noise of a multispectral image can be efficiently removed due to the retaining of proper singular values that can usually represent original image. Turn to PSNR values, it is clearly concluded that the proposed GA-SVD method show almost as much high performance as the traditional real SVD method.

Table 2.1 PSNR values (dB) obtained by different denoising methods at each noisy level

δ	methods	balloon	pepper	feather	ball	flower	toy
15	SVD	23.16	20.92	18.99	18.99	19.09	23.78
	GA-SVD	30.65	30.33	30.11	30.11	29.07	32.46
25	SVD	22.84	20.07	20.07	17.93	18.80	23.67
	GA-SVD	29.91	29.25	29.25	29.46	28.40	31.44
35	SVD	20.63	19.91	19.91	17.58	18.66	23.20
	GA-SVD	28.61	27.94	27.94	28.22	27.30	29.76
45	SVD	19.05	19.23	19.23	17.06	18.20	22.92
	GA-SVD	26.88	26.33	26.33	26.49	25.85	27.68

2.6 Conclusions

In this chapter, we proposed a novel singular value decomposition algorithm GA-SVD based on geometric algebra for multispectral image by converting it to the real SVD of its equivalent real representation. GA-SVD represents multispectral images as GA multivectors to capture the relationship of these spectral channels. The GA-SVD and SVD algorithms are implemented in the application of compression and denoising of multispectral images. And the comparison results show that GA-SVD is better in preserving detailed information. Many SVD-based image processing methods can be extended for multispectral images, such as detection, enhancement, classification, object tracking and so on. Consequently, the proposed GA-SVD is an efficient and homogeneous tool in multispectral image analysis.

References

[1] Said B, Hadjidj R, Melkemi K E, et al. Multispectral image denoising with optimized vector non-local mean filter. Digital Signal Processing, 2016, 58(C): 115-126.

[2] Tzoumas S, Rosenthal A, Lutzweiler C, et al. Spatio spectral denoising framework for multispectral optoacoustic imaging based on sparse signal representation. Medical Physics, 2014, 41(11): 113301.

[3] Xie Q, Zhao Q, Meng D, et al. Multispectral images denoising by intrinsic tensor sparsity regularization. Computer Vision & Pattern Recognition, 2016: 1692-1700.

[4] Valsesia D, Magli E. A novel rate control algorithm for onboard predictive coding of multispectral and hyperspectral images. IEEE Transactions on Geoscience and Remote Sensing, 2014, 52(10): 6341-6355.

[5] Valsesia P, Jeyakumar S, Nithiyanandam N. Hyperspectral image compression based on non-iterative matrix factorization. 2013 IEEE International Conference on Computational Intelligence and Computing Research (ICCIC). Piscataway: IEEE, 2013.

[6] Kawakami R, Matsushita Y, Wright J, et al. High-resolution hyperspectral imaging via matrix factorization. Computer Vision and Pattern Recognition, 2011, 42: 2329-2336.

[7] Chen A. The inpainting of hyperspectral images: A survey and adaptation to hyperspectral data. Image and Signal Processing for Remote Sensing XVIII, 2011, 8537: 1-8.

[8] Nguyen H V, Banerjee A, Chellappa R. Tracking via object reflectance using a hyperspectral video camera. IEEE Computer Society Conference on Computer Vision and Pattern Recognition – Workshops IEEE, 2011: 44-51.

[9] Shih Y T, Chien C S, Chuang C Y. An adaptive parameterized block-based singular value decomposition for image de-noising and compression. Applied Mathematics & Computation, 2012, 218(8): 10370-10385.

[10] Andrews H C, Patterson C L. Singular value decompositions and digital image processing. IEEE Transactions on Acoustics, Speech and Signal Processing, 1976, 1(24): 26-53.

[11] Huang J, Xie J, Li F, et al. Compound de-noising method based on discrete stationary wavelet transform and optimized SVD. Piezoelectrics & Acoustooptics, 2013, 35(3): 448-451.

[12] Chen L, Zhang X, Shen Y J, et al. Fault diagnosis of rolling bearings based on SVD denoising and blind signals separation. Journal of Vibration & Shock, 2012.

[13] Zhao X Z, Yan B, Chen T J. Principle of singularity detection based on SVD and its application. Journal of Vibration & Shock, 2008.

[14] Zhang F Z. Quaternion and matrices of quaternions. Linear Algebra & Its Applications, 1997, 251(2): 21-57.

[15] Bihan N L, Mars J. Singular value decomposition of quaternion matrices: A new tool for vector-sensor signal processing. Signal Processing, 2004, 84(7): 1177-1199.

[16] Lei Y J, Yan Y U, Zhou J L, et al. Quaternion singular value decomposition approach to color image de-noising. Journal of Sichuan University, 2008, 44(6): 131-136.

[17] Franchini S, Gentile A, Sorbello F, et al. Conformal ALU: A conformal geometric algebra coprocessor for medical image processing. IEEE Transactions on Computers, 2015, 64(4): 955-970.

[18] Su H, Bo Z. Conformal geometric algebra based band selection and classification for hyper spectral imagery. 2016 8th Workshop on Hyperspectral Image and Signal Processing: Evolution in Remote Sensing (WHISPERS), Los Angeles, CA, 2016: 1-4.

[19] Pham M T, Yoshikawa T, Furuhashi T, et al. Robust feature extractions from geometric data using geometric algebra. IEEE International Conference on Systems Man and Cybernetics, 2009: 529-533.

[20] Ebling J, Scheuermann G. Clifford Fourier transform on vector fields. IEEE Transactions on Visualization & Computer Graphics, 2005, 11(4): 469.

[21] Wang R, Zhou Y X, Jin Y L, et al. Sparse fast Clifford Fourier transform. Frontiers of Information Technology & Electronic Engineering, 2017, 18(8): 1131-1141.

[22] Li Y, Liu W, Li X, et al. GA-SIFT: A new scale invariant feature transform for multispectral image using geometric algebra. Information Sciences, 2014: 559-572.

[23] Gunn C. A new approach to euclidean plane geometry based on projective geometric algebra. Mathematics, 2015, 704(4): 41-58.

[24] Hitzer E, Nitta T, Kuroe Y. Applications of Clifford's geometric algebra. Clifford Algebras, 2013, 23(2): 377-404.

[25] Zhou S L, Valchev D G, Dinovitser A, et al. Dispersion-independent terahertz classification based on geometric algebra for substance detection. 2016 41st International Conference on Infrared, Millimeter, and Terahertz Waves (IRMMW-THz), Copenhagen, 2016: 1, 2.

[26] Stewart G W. On the early history of the singular value decomposition. Society for Industrial and Applied Mathematics, 1993.

[27] Lathauwer D, Moor D, Vandewalle J. A multilinear sigular value decomposition. SIAM Journal on Matriz Analysis and Applications, 2000, 21(4): 1253-1278.

[28] Li Y, Wei M, Zhang F, et al. A fast structure-preserving method for computing the singular value decomposition of quaternion matrices. Applied Mathematics & Computation, 2014, 235(4): 157-167.

[29] Li Y, Wei M, Zhang F, et al. Real structure-preserving algorithms of Householder based transformations for quaternion matrices. Journal of Computational & Applied Mathematics, 2016, 305: 82-91.

[30] Yasuma F, Mitsunaga T, Iso D, et al. Generalized assorted pixel camera: postcapture control of resolution, dynamic range, and spectrum. IEEE Trans. Image Processing, 2010, 19(9): 2241-2253, 2010.

Chapter 3

Multivector Sparse Representation for Multispectral Images Using Geometric Algebra

3.1 Introduction

Recently, multispectral images have been widely applied in more and more areas, e.g., medical image analysis, military defense, environmental monitoring, space remote sensing, geological exploration, digital photography, etc.[1−3], due to the large amount of information it contains. Combined with spatial and spectral information, multispectral images specifically consist of visible light, infrared ray, ultraviolet radiation, millimeter wave or X-ray. That is to say, a multispectral image is formed with multiple scalar images, in which each image is usually regarded as a band or a channel, at specific frequencies across the electromagnetic spectrum. Substantially, the performance of multiple computer vision tasks, such as tracking, and inpainting[4,5], have been dramatically enhanced due to such comprehensive knowledge representation capability of multispectral images. Therefore, the processing techniques for multispectral images have become quite emerging topics, which bring both opportunities and challenges[6−8].

In image processing and analysis domain, sparse representation[9,10] has occupied an important place, and has witnessed great success in variety of tasks, e.g., natural image denoising[11,12], reconstruction[12], classification[13,14], detection[15], and restoration[16]. The K-SVD algorithm[11] was utilized to show great performance in a wide range of gray-scale image processing, but it leaded to some unsatisfying results for color image due to the loss of relationship between RGB channels. Mairal et al.[16] provided a simple concatenation of the RGB values of a color image to a single vector and training on those directly, which gives already better results than performing each channel separately. However, such a process produces false colors and artifacts. Xu et al.[17] provided an efficient color sparse representation model on account of the compatibility between quaternion matrix and color image to avoid losing the correlation of three color channels. Wang et al.[18] applied geometric algebra to map the color image into a high dimensional space to color image analysis and achieved high performance. To address the problem of high computational complexity owing to non-commutative multiplication, Shen et al.[19] presented a novel theory of reduced geometric algebra (RGA) with commutative multiplication rules and a novel vector-valued sparse representation model for color image using RGA. Shen et al.[20] then proposed a new singular value decomposition algorithm for octonion signals, which can be simply regarded as seven-channel multispectral images. However, existing sparse models are mostly developed for gray or 3-channel color images[17,21,22], and there are very few works focusing on the sparse representation of multichannel signals, typically as multispectral images.

Fortunately, geometric algebra (GA) provides a powerful computing framework in signal and image processing and it gives a formidable way for multi-dimensional signals[23–25]. Ebling et al.[26] extended the Fourier transform to (multi) vector fields using GA and verify the generalizations of correlation, convolution and derivative theorems. Wang et al.[27] created a sparse fast Clifford Fourier transform based on GA, which showed great improvement in computing performance over scalar and vector fields. Li et al.[28] put forward a new SIFT algorithmic framework based on GA for multispectral images processing and analysis.

Inspired by the useful information contained in multispectral images, the recent progress of sparse representation models and the advantages of GA theory in various fields of image processing, we present a novel multivector sparse representation model, which adopts GA framework to leverage the multispectral information. Due to the utilization of the GA theory, both spatial and spectral information of multispectral images can be completely retained, and the inherent spectral structures in multispectral images can be preserved successfully during sparse reconstruction without losing the relationship of spectral channels. Besides, in terms of training process, we correspondingly design the dictionary learning algorithm based on GA, namely K-GASVD (K-means clustering for singular value decomposition based on GA) to obtain a trained dictionary that contains the richest spectral information. In Summary, the GA dictionary converts the channel images to a subspace in a uniform way, and consequently the inherent spectral structures can be successfully preserved during vector reconstruction.

The major contribution of this chapter lies in the following aspects. Firstly, a multispectral image is successfully represented as a GA multivector, in which each spectral channel is mapped to each blade of GA. Secondly, we propose a novel and efficient multivector sparse representation model based on GA on account of the compatibility between a GA multivector and a multispectral image to avoid losing the correlation of spectral channels. Thirdly, we develop an effective dictionary learning method based on GA, namely K-GASVD, which has the ability to update coefficient simultaneously instead of fixing the coefficient vector during dictionary learning. Extensive experiments of multispectral images reconstruction and denoising indicate that superior performance is achieved with the proposed sparse model.

The remainder of this chapter is organized as follows. Section 3.2 reviews the existing sparse models and basis of geometric algebra (GA) briefly. The proposed sparse model for multispectral images is illustrated in Section 3.3. GA based SVD and dictionary training methods are described in Section 3.4. Reconstruction and denoising experiments for multispectral images using our proposed model are presented in Section 3.5, and the conclusion of this chapter is given in Section 3.6.

3.2 Related Work

3.2.1 Review of Current Sparse Representation Models

As a powerful and vibrant method to represent images utilizing redundant dictionaries, image sparse representations that mostly focused on patch processing, have enjoyed great popularity as a means for image processing and analysis. It is well known that existing sparse models have been very successful in handling single channel gray-scale images as well as 3-channel color images, in which every color channel is associated with each other.

Sparse model based on K-SVD method[11] has achieved great performance in gray-scale image domain, and can be formulated by

$$\min_{a_s} \|a_s\|_0, \quad \text{s.t.} \quad f_s = D_s a_s \tag{3.1}$$

where $f_s \in \mathbb{R}^n$ is the gray-scale image patch, $D_s \in \mathbb{R}^{n\times m}$ is the corresponding dictionary, and $a_s \in \mathbb{R}^m$ is the assumed sparse coefficient vector. Nevertheless, for color image, it just regards RGB channels as three independent "gray-scale" images and processes them in a monochrome way, which loses the inter-correlation among the three-color channels leading to some unsatisfying results.

To further avoid the loss of relationship among channels in a color image, the approach put forward in Ref.[16] concatenated the three channels into a single channel to train a common dictionary and share a common coefficient vector, the concatenation model is described as follows

$$[y_r^{\mathrm{T}} \quad y_g^{\mathrm{T}} \quad y_b^{\mathrm{T}}]^{\mathrm{T}} = [D_r^{\mathrm{T}} \quad D_g^{\mathrm{T}} \quad D_b^{\mathrm{T}}]^{\mathrm{T}} a \tag{3.2}$$

Though, the concatenation model in (3.2) partly addressed the loss of inherent color structure compared to (3.1), it inevitably produced some unsatisfying results such as false colors and artifacts.

When dealing with color image, the most urgent problem needed to be addressed is to avoid the loss of inherent relationship of RGB channels. Wang et al.[18] applied geometric algebra to denote the structure of color information and map the color image into a high-dimensional space, and the model has been, until very recently, regarded as a powerful model for color image processing. Each color image patch can be represented as $f = 0 + f_1e_1 + f_2e_2 + f_{12}e_{12} \in (\mathbb{G}_2)^N$ while applying GA, and the sparse representation model in $\mathbb{G}_2$ has been given as

$$\min_g \|g\|_0, \quad \text{s.t.} \quad f = Dg \tag{3.3}$$

where $D = 0 + D_1e_1 + D_2e_2 + D_{12}e_{12} \in (\mathbb{G}_2)^{N\times M}$ is a dictionary in $\mathbb{G}_2$, $g = 0 + g_1e_1 + g_2e_2 + g_{12}e_{12} \in (\mathbb{G}_2)^M$ is a corresponding sparse coefficient vector, with its components g_1; g_2 and $g_{12} \in \mathbb{R}^M$. Here, $\mathbb{G}_2$ denotes the 2-dimensional GA space.

Compared with the K-SVD based sparse model in (3.1), the sparse representation model in $\mathbb{G}_2$ deals with three color channels information in a unified way to imitate the process of human visual perception. The coefficient matrix captures both the relationship of color channels and the orthogonal property successfully. The obvious advantage is that it maintains the inherent structures of RGB channels during image reconstruction.

In retrospect, the performance improvements of these recent attempts have been ascribed to their use of algebraic frameworks and, until now, very few researches have been conducted in handling high dimensional data, typically as multispectral images. Then, in this chapter, we pay more attention to retain the correlation of each spectral channel by representing the multispectral images and subsequently propose a multivector sparse representation model based on geometric algebra.

3.2.2 Representation Models for Multispectral Images

Since multispectral images have recently attracted much attention and there have emerged many representation method. Huang et al.[29] presented a novel spatial and spectral fusion model (SASFM) that uses sparse matrix factorization to fuse remote sensing imagery with different spatial and spectral properties. Lanaras et al.[30] jointly processed high spectral and high geometric resolution images and exploited their synergies to generate a fused image of high spectral and geometric resolution, and improve (linear) spectral unmixing of hyperspectral endmembers at subpixel level, which is the pixel size of the hyperspectral image. Dian et al.[31] proposed a novel hyperspectral image (HIS) super-resolution method based on non-local sparse tensor factorization (called as the NLSTF). The sparse tensor factorization can directly decompose each cube of the HSI as a sparse core tensor and dictionaries of three modes, which reformulates the HIS super-resolution problem as the estimation of sparse core tensor and dictionaries for each cube. Zhang et al.[32] proposed a new low-resolution HS (LRHS) and high-resolution MS (HRMS) image fusion method based on spatial-spectral-graph- regularized low-rank tensor decomposition (SSGLRTD) to effectively preserve spatial-spectral structures in HRHS images.

3.2.3 The Basics of Geometric Algebra

William K. Clifford introduced geometric algebra abbreviated as GA, also called Clifford algebra, which provides such a coordinate-free framework to make the computation efficiently[23−25]. It completes the constructions and modellings in a coordinate-free way and also has revealed wide applications especially when applying to computer vision tasks[26−28].

Suppose $\mathbb{G}_n$ is n-dimensional GA with an orthonormal basis of vectors $\{e_i\}, i = 1, \cdots, n$, which leads to a basis

$$\{1, \{e_i\}, \{e_i e_j\}, \cdots, \{e_1 e_2 \cdots e_n\}\} \tag{3.4}$$

Specifically, $\mathbb{G}_n$ can be represented by $\mathbb{G}_{p,q}$, where $n = p + q$ and p, q are the number of vectors with positive square and negative square in the basis of the space, respectively.

That is

$$e_i^2 = \begin{cases} 1, & 1 \leqslant i \leqslant p \\ -1, & p+1 \leqslant i \leqslant n \end{cases} \tag{3.5}$$

In this chapter, we focus on $\mathbb{G}_n$, in which $q = 0$. Generally, the geometric product of two basis is anti-commutative, and

$$e_i e_j = e_{ij} = -e_j e_i = -e_{ji}, \quad i, j = 1, \cdots, n, \quad i \neq j \tag{3.6}$$

$$e_i^2 = 1, \quad i = 1, \cdots, n \tag{3.7}$$

$$e_i e_{ij} = e_i e_i e_j = e_j, \quad i, j = 1, \cdots, n, \quad i \neq j \tag{3.8}$$

Given two vectors w and z in $\mathbb{G}_n$, the geometric product is given by

$$wz = w \cdot z + w \wedge z \tag{3.9}$$

where $w \cdot z$ denotes the inner product and $w \wedge z$ is the outer product. Since the vectors are orthogonal, then $e_i e_j = e_i \cdot e_j + e_i \wedge e_j = e_i \wedge e_j$.

Take $\mathbb{G}_3$ for instance, the orthogonal bases are constructed by vectors with $2^3 = 8$ grades, which is given by: $\{1, \{e_1, e_2, e_3\}, \{e_1e_2, e_2e_3, e_3e_1\}, \{e_1e_2e_3\}\}$ and a simpler form: $\{1, e_1, e_2, e_3, e_{12}, e_{23}, e_{31}, e_{123}\}$. For a 2^n-dimensional signal, only basis is needed. Generally, the outer product of k vectors is called a k-blade. The number k is called the grade of the blade.

In GA space, multivectors, which stand for the extension of vectors in higher dimensions, are the basic elements. Any multivector $M \in \mathbb{G}_n$ is described by

$$M = E_0 + \sum_{1 \leqslant i \leqslant n} E_i(M)e_i + \sum_{1 \leqslant i < j \leqslant n} E_{ij}(M)e_{ij} + \cdots + E_{1\cdots n}(M)e_{1\cdots n} \tag{3.10}$$

where $E(M) \in \mathbb{R}$.

3.3 The Multivector Sparse Represention Model for Multispectral Images

In this section, we utilize geometric algebra (GA) to extend the sparse model in (3.3) for color image to multispectral images.

3.3.1 Representation of Multispectral Images Using GA

Each pixel of a multispectral image I can be denoted in G_n as follows.

$$I(x, y) = 0 + \sum_{1 \leqslant i \leqslant n} I_i(x, y)e_i + \sum_{1 \leqslant i < j \leqslant n} I_{ij}(x, y)e_{ij} + \cdots + I_{1\cdots n}(x, y)e_{1\cdots n} \tag{3.11}$$

where $I_i(x, y), I_{ij}(x, y), I_{1\cdots n}(x, y)$ are the spectral components of multispectral image pixel $I(x, y)$ at position (x, y) respectively.

Specifically, all the spectral channels of the multispectral image are assigned to the basis $\{1, \{e_i\}, \{e_i e_j\}, \{e_1 e_2 \cdots e_n\}\}$ of the 2^n-dimensional GA ($\mathbb{G}_n$) respectively, and a GA multivector is used to rewrite the multispectral images pixel $I(x, y)$ in GA space.

Thus, a multispectral image $F \in (\mathbb{G}_n)^{M \times N}$ with M rows and N columns can be represented by a GA multivector

$$F = 0 + \sum_{1 \leqslant i \leqslant n} E_i(F)e_i + \sum_{1 \leqslant i < j \leqslant n} E_{ij}(F)e_{ij} + \cdots + E_{1\cdots n}(F)e_{1\cdots n}, E(F) \in \mathbb{R}^{M \times N} \tag{3.12}$$

Recently, the GA representation of multispectral images have attracted great attention. Since our work is not the first one to represent multispectral images using GA, the representation method based on GA introduced in this chapter has some differences from the paper by Li et al.[28]. In the paper by Li et al., it only uses 1-blade to represent a multispectral image (see Eq.(3.14)) and ignores all the outer product parts (it says "There are two multivector in the result of the Eq.(3.27), and the second multivector cannot be used in the image processing simply."). In our chapter, we reserve all the outer product parts, which brings the relationship of components in GA space into correspondence with the correlation of spectral channels in a multispectral image, and can be considered as a better and more comprehensive representation of a multispectral image. It can be applied in studying the problems of multispectral images processing due to its capability to treat the spectral channels holistically without losing the relationship of the spectral channels.

3.3.2 GA-Multivector Sparse Representation Model for Multispectral Images

To take the correlation of the multispectral channels into consideration, we put forward a novel multivector sparse representation model using geometric algebra for multispectral images in this chapter. For each patch $F \in (\mathbb{G}_n)^{\sqrt{N}\times\sqrt{N}}$ of the multispectral image $F \in (\mathbb{G}_n)^{N\times K}$, where N is the size of the multispectral image and K denotes the number of image patches. Then we transfer it to a vector $f \in (\mathbb{G}_n)^N$ with the length of N as follows

$$f = 0 + \sum_{1\leqslant i\leqslant n} E_i(f)e_i + \sum_{1\leqslant i<j\leqslant n} E_{ij}(f)e_{ij} + \cdots + E_{1\cdots n}(f)e_{1\cdots n}, E(f) \in \mathbb{R}^N \quad (3.13)$$

Then, we define our proposed multivector sparse representation model as follows

$$\min_a \|a\|_0, \quad \text{s.t.} \quad f = Da \quad (3.14)$$

where $D = \begin{pmatrix} E_0(D) + \sum_{1\leqslant i\leqslant n} E_i(D)e_i + \sum_{1\leqslant i<j\leqslant n} E_{ij}(D)e_{ij} \\ + \cdots + E_{1\cdots n}(D)e_{1\cdots n} \in (\mathbb{G}_n)^{N\times M} \end{pmatrix}$ is a GA dictionary consisting of M atoms and $E(D) \in \mathbb{R}^{N\times M}$.

$$a = \begin{pmatrix} E_0(a) + \sum_{1\leqslant i\leqslant n} E_i(a)e_i + \sum_{1\leqslant i<j\leqslant n} E_{ij}(a)e_{ij} \\ + \cdots + E_{1\cdots n}(a)e_{1\cdots n} \in (\mathbb{G}_n)^M \end{pmatrix}$$

is a GA coefficient vector and $E(a) \in \mathbb{R}^M$. The objective function $\|a\|_o$ is resolved to collect all the number of nonzero components in the coefficient vector above.

The generalized form of representation model for multispectral images is derived in the Appendix A.

Specially, for a multispectral image $F' \in (\mathbb{G}_2)^{N\times K}$ with 3 channels, which can be represented in $\mathbb{G}_2$ space, and each patch $F \in (\mathbb{G}_2)^{\sqrt{N}\times\sqrt{N}}$ can be transferred to a vector $f' \in (\mathbb{G}_2)^N$ with the length of N as follows

$$f' = 0 + E_1(f')e_1 + E_2(f')e_2 + E_{12}(f')e_{12} \in (\mathbb{G}_2)^N \tag{3.15}$$

Then, the proposed model in $\mathbb{G}_2$ space for 3 channels color image is defined as follows

$$\min_{a'} \|a'\|_0, \quad \text{s.t.} \quad f' = D'a' \tag{3.16}$$

where $D' = E_0(D') + E_1(D')e_1 + E_2(D')e_2 + E_{12}(D')e_{12}, D' \in (\mathbb{G}_2)^{N\times M}$ is defined as a GA dictionary with M atoms and a GA coefficient vector is shown as $a' = E_0(a') + E_1(a')e_1 + E_2(a')e_2 + E_{12}(a')e_{12}$.

Similarly, $\|a\|_0$ is utilized to count the number of all the non-zero components in GA coefficient vector.

The generalized form of 3 channels color image sparse model can then be obtained as follows

$$[0 \quad E_1(f') \quad E_2(f') \quad E_{12}(f')] = [E_0(D') \quad E_1(D') \quad E_2(D') \quad E_{12}(D')]\,(a') \tag{3.17}$$

Since

$$\begin{aligned}
f' = D'a' \Leftrightarrow\ & 0 + E_1(f')e_1 + E_2(f')e_2 + E_{12}(f')e_{12} \\
&= (E_0(D') + E_1(D')e_1 + E_2(D')e_2 + E_{12}(D')e_{12}) \\
&\quad (E_0(a') + E_1(a')e_1 + E_2(a')e_2 + E_{12}(a')e_{12}) \\
&= (E_0(D')E_0(a') + E_1(D')E_1(a') + E_2(D')E_2(a') - E_{12}(D')E_{12}(a')) \\
&\quad + (E_0(D')E_1(a') + E_1(D')E_0(a') - E_2(D')E_{12}(a') + E_{12}(D')E_2(a'))\,e_1 \\
&\quad + (E_0(D')E_2(a') + E_1(D')E_{12}(a') + E_2(D')E_0(a') - E_{12}(D')E_1(a'))\,e_2 \\
&\quad + (E_0(D')E_{12}(a') + E_1(D')E_2(a') - E_2(D')E_1(a') + E_{12}(D')E_0(a'))\,e_{12}
\end{aligned} \tag{3.18}$$

Therefore, we can obtain

$$(a') = \begin{bmatrix} E_0(a') & E_1(a') & E_2(a') & -E_{12}(a') \\ E_1(a') & E_0(a') & -E_{12}(a') & E_2(a') \\ E_2(a') & E_{12}(a') & E_0(a') & -E_1(a') \\ E_{12}(a') & E_2(a') & -E_1(a') & E_0(a') \end{bmatrix} \tag{3.19}$$

3.4 GA-based Dictionary Training

In this section, we extend the singular value decomposition (SVD) to n-dimensional space using GA, and present the dictionary-learning problems.

Similar to the SVD of real data[33−35], for a GA multivector $F \in (\mathbb{G}_n)^{M\times N}$ of a multispectral image, then there exist two GA unitary matrix $U \in (\mathbb{G}_n)^{M\times M}$ and $V \in (\mathbb{G}_n)^{N\times N}$ such that

$$U^H F V = \begin{bmatrix} \Sigma' & 0 \\ 0 & 0 \end{bmatrix} \equiv \Sigma \in \mathbb{R}^{M\times N} \tag{3.20}$$

where $\Sigma' = \text{diag}\{\sigma_1, \sigma_2, \cdots, \sigma_P\} \in \mathbb{R}^{P\times P}$, σ_i are the positive singular values of F, $\sigma_1 \geqslant \sigma_2 \geqslant \cdots \geqslant \sigma_P$ and $P = \min\{M, N\}$ and H represents the conjugate transpose operation.

That is, for any $M \times N$ GA multivector F, its SVD is obtained by

$$F = U\Sigma V^H = U \begin{bmatrix} \Sigma' & 0 \\ 0 & 0 \end{bmatrix} V^H \tag{3.21}$$

where $UU^H = I^{M\times M}, VV^H = I^{N\times N}$, I is an identity matrix.

We can perform the GA-SVD with different kinds of algorithms.

Suppose

$$F = E_0(F) + \sum_{1\leqslant i\leqslant n} E_i(F)e_i + \sum_{1\leqslant i<j\leqslant n} E_{ij}(F)e_{ij} + \cdots + E_{1\cdots n}(F)e_{1\cdots n} \in (\mathbb{G}_n)^{M\times N}$$

where $E(F) \in \mathbb{R}^{M\times N}$, its real representation F_R can be defined as (3.22).

$$F_R = \begin{bmatrix} E_0(F) & \cdots & E_t(F) & \cdots \\ \vdots & & \vdots & \\ E_t(F) & \cdots & E_0(F) & \cdots \\ \vdots & & \vdots & \\ E_{ts}(F) & \cdots & -E_s(F) & \cdots \\ \vdots & & \vdots & \\ E_{1\cdots n}(F) & \cdots & (-1)^{(n-t)}E_y(F) & \cdots \end{bmatrix}$$

$$\begin{array}{ccc} -E_{ts}(F) & \cdots & (-1)^{n(n-1)/2}E_{1\cdots n}(F) \\ \vdots & & \vdots \\ -E_s(F) & \cdots & (-1)^{(t-1)+(n(n-1)/2)}E_{1\cdots n}(F) \\ \vdots & & \vdots \\ E_0(F) & \cdots & (-1)^{(t-1)+(s-2)+(n(n-1)/2)}E_x(F) \\ \vdots & & \vdots \\ (-1)^{(2n-t-s+1)}E_x(F) & \cdots & E_0(F) \end{array}\Bigg] \tag{3.22}$$

where

$$x = 1\cdots(t-1)(t+1)\cdots(s-1)(s+1)\cdots n; y = 1\cdots(t-1)(t+1)\cdots n;$$

$$t = 1\cdots n; \quad i \neq t, \quad t \neq s, \quad j \neq t, \quad j \neq s; \quad j > i, \quad s > t;$$

$$E_{ij} = E_{ji} \quad \text{and} \quad F_R \in \mathbb{R}^{2^n M \times 2^n N}.$$

Then using the isomorphism between $(\mathbb{G}_n)^{M\times N}$ and $(\mathbb{R})^{2^n M\times 2^n N}$, the GASVD of F can be performed by employing the conventional real SVD algorithm to F_R. We define the relationship between the GASVD of a GA multivector F and the SVD of its equivalent real matrix F_R as follows

1) $\Sigma' = \text{row}_{\text{odd}}^{2^{n-1}}(\text{col}_{\text{odd}}^{2^{n-1}}(\Sigma'_R))$.

2) $U = \left(E_0(U) + \sum\limits_{1\leqslant i\leqslant n} E_i(U)e_i + \sum\limits_{1\leqslant i<j\leqslant n} E_{ij}(U)e_{ij} + \cdots + E_{1\cdots n}(U)e_{1\cdots n}\right)$,

$U_R = [\,(U_R)_0 \quad \cdots \quad (U_R)_i \quad \cdots \quad (U_R)_{ij} \quad \cdots \quad (U_R)_{1\cdots n}\,]^{\mathrm{T}}$,

$$U = \left((U_R)_0 - \sum_{1\leqslant i\leqslant n}(U_R)_i e_i - \sum_{1\leqslant i<j\leqslant n} E_{ij}(U)e_{ij}\right) + \cdots + E_{1\cdots n}(U)e_{1\cdots n}.$$

3) $V = \left(E_0(V) + \sum\limits_{1\leqslant i\leqslant n} E_i(V)e_i + \sum\limits_{1\leqslant i<j\leqslant n} E_{ij}(V)e_{ij} + \cdots + E_{1\cdots n}(V)e_{1\cdots n}\right)$,

$V_R = [\,(V_R)_0 \quad \cdots \quad (V_R)_i \quad \cdots \quad (V_R)_{ij} \quad \cdots \quad (V_R)_{1\cdots n}\,]^{\mathrm{T}}$,

then $V = \left((V_R)_0 - \sum\limits_{1\leqslant i\leqslant n}(V_R)_i e_i - \sum\limits_{1\leqslant i,j\leqslant n}(V_R)_{ij}e_{ij} - \cdots - (V_R)_{1\cdots n}(U)e_{1\cdots n}\right)$

where $1 \leqslant i < j \leqslant n$, $\text{row}_{\text{odd}}^{2^{n-1}}(P)$ and $\text{col}_{\text{odd}}^{2^{n-1}}(P)$ denote the odd rows and odd columns extracted from matrix P for $2n-1$ times respectively.

3.4.1 GA Dictionary Training Analysis

With the unknown variables, dictionary and coefficients, the dictionary training process using GA can be regarded as an extension of the sparse model put forward in Eq.(3.16), and can be formulated as

$$\{\hat{D} \quad \hat{A}\} = \arg\min_{D,A} \|F - DA\|_F^2 + \gamma \|A\|_0 \tag{3.23}$$

where $F = \{f_i, 1 \leqslant i \leqslant K\} \in (\mathbb{G}_n)^{N\times K}$ is a series of the sample image patches, $D = \{d_i, 1 \leqslant i \leqslant M\} \in (\mathbb{G}_n)^{N\times M}$ is the GA dictionary with M atoms, $A = \{a_i, 1 \leqslant i \leqslant K\} \in (\mathbb{G}_n)^{M\times K}$ is the sparse coefficient matrix and $\|A\|_0$ counts the nonzero elements of the columns in A.

K-SVD[11] is a kind of fast and efficient sparse dictionary learning algorithm, whose key problem is how to learn sparse dictionary adaptively from the training data set, and it performs well for gray-scale image processing. However, when adapted to multispectral images, the main challenge is to capture the loss of relationship among different spectral channels during image reconstruction. In this chapter, a dictionary training algorithm based on GA, called K-GASVD, is introduced as an extension of K-SVD[11] to cater to the multispectral images vectorization. Particularly, the GA based dictionary training process contains two major steps, i.e., sparse coding stage and dictionary updating stage.

After extending to multispectral images processing, the orthogonal matching pursuit algorithm using GA (GAOMP) is designed to handle the sparse representation of GA multivector. The GAOMP algorithm deals with the problem of decomposing signal $F \in (\mathbb{G}_n)^{N\times K}$ on a GA dictionary $D \in (\mathbb{G}_n)^{N\times M}$ such that

$$A = \arg\min_A \|F - DA\|_2^2, \quad \text{s.t.} \quad \|A\| \leqslant T \tag{3.24}$$

where $A \in (\mathbb{G}_n)^{M\times K}$ denotes the sparse coefficient vector and $\|A\| \leqslant T$ is the stopping criteria. Such operation helps to alleviate the problem of l_0-norm sparse coding that is usually seen as a NP-hard problem through assigning the maximum number of non-zero coefficients of each signal.

The details of GAOMP algorithm for each multispectral image patch are given in Table 3.1.

Once the sparse coding solution is obtained, the dictionary and sparse representation coefficient are updated simultaneously using GASVD. The details of K-GASVD algorithm is shown in Table 3.2.

Table 3.1 The details of GA-based orthogonal matching pursuit algorithm

Input: multispectral image patch: $f \in (\mathbb{G}_n)^N$; GA-based dictionary: $D = \{d_j, 1 \leqslant j \leqslant M \in (\mathbb{G}_n)^{N\times M}\}$; sparsity level: T.
Output: sparse coefficient: $a \in (\mathbb{G}_n)^M$.
1. Initialization: $i = 1$, $a^0 = 0$, $\varepsilon^0 = f - Da^0 = f, \Phi^0 = \varnothing$.
2. Set: $i = i + 1$.
3. Select $j_0 \forall j$, s.t. $\min_k \left\|kd_{j_0} - \varepsilon^{i-1}\right\|_2 \leqslant \min_k \left\|kd_j - \varepsilon^{i-1}\right\|_2$.
4. Set: $\Phi_i = \Phi^{i-1} \cup \{j_0\}$.
5. Calculate: $a^i = \arg\min_a \|D_a - f\|_2$, s.t. $a = \Phi^i$.
6. Calculate: $\varepsilon^i = f - Da^i$.
7. When $i > T$, stop the iteration.

Table 3.2 The details of GA-based dictionary learning using K-GASVD method

Input: multispectral image patches: $F = \{f_i, 1 \leqslant i \leqslant K\} \in (\mathbb{G}_n)^{N\times K}$; GA-based dictionary: $D = \{d_i, 1 \leqslant i \leqslant M\} \in (\mathbb{G}_n)^{N\times M}$; the number of iterations: J.
Output: sparse coefficient: $A = \{a_i, 1 \leqslant i \leqslant K\} \in (\mathbb{G}_n)^{M\times K}$; dictionary after training: $D \in (\mathbb{G}_n)^{N\times K}$.
1. Initialization: $j = 0$, initialize the GA-based dictionary $D \in (\mathbb{G}_n)^{N\times M}$ as random M samples from $F \in (\mathbb{G}_n)^{N\times K}$.
2. When $j < J$, repeat
(1) Sparse coding: compute sparse representation coefficient $a \in (\mathbb{G}_n)^M$ of $F \in (\mathbb{G}_n)^{N\times K}$ over $D \in (\mathbb{G}_n)^{N\times M}$ using GAOMP algorithm.
(2) Dictionary updating: update each dictionary atom $d_k \in (\mathbb{G}_n)^N$ in D^{J-1}.
(i) Define the set of patches using the k-th dictionary atom $d_k \in (\mathbb{G}_n)^N$ as $\Omega_k = \{i \mid 1 \leqslant i \leqslant k, A(k,i) \neq 0\}$.
(ii) Calculate the sparse error matrix E_k without using d_k. $E_k = F - \sum\limits_{i\neq k} D_i a^i$.
(iii) Select columns of E_k corresponding to Ω_k to E_K^Ω. $E_K^\Omega = E_K(:,i)\|_{i=\Omega_k}$.
(iv) Perform the SVD of E_k using GASVD: $E_K^\Omega = U\Sigma V^H$ and update dk as U (:,1), set $a_K^\Omega = \Omega_K^{\mathrm{T}} a^k$ to be $V^H(:,1)\Sigma(1,1)$.

3.4.2 Further Analysis

It is obvious that the introduced GAOMP algorithm in Table 3.1 is also a greedy algorithm, which mainly focuses on selecting an atom that best matches the signal from the dictionary according to certain criteria during each iteration to sparsely approximate the original signal. Similar to the traditional OMP algorithm for scalar data[36], the introduced GAOMP algorithm is designed for multi-dimensional signals, typically for multispectral images, based on GA. At each step, the best atom which denotes the maximum of its inner product with the residual, is selected from GA dictionary, and then the residual is updated by conducting an orthogonal projection of the signal approximated onto the vectorial space which is generated by the chosen atoms formerly. Clearly, this orthogonalization shows its absolute importance since it makes the convergence of this greedy algorithm more stably and faster.

For the introduced GAOMP algorithm, because of the introduction of GA, the result of inner product includes all the components of multispectral images, rather than a single component of original grayscale image. Therefore, during the iterations, the sparsity of all the spectral channels rather than a single channel is always guaranteed. So, all the components of multispectral images are processed integrally and monolithically while searching for the nearest atom.

Moreover, the convergence of the provided K-GASVD is similar to that of K-SVD, since it applies the same framework as the traditional K-SVD[9,11,13]. Fig.3.1 shows the trained dictionaries based on K-SVD and K-GASVD methods. We observe that the learned dictionary using K-SVD method shows high tendency to be monochromatic. Since the K-SVD algorithm cannot perfectly handle the rich spectral components of the original multispectral images, which produces a large number of grayscale blocks in the training dictionary, resulting in some unsatisfying results of the reconstructed image, such as color is not rich enough and the color saturation is significantly reduced. In contrast, the learned dictionary based on K-GASVD

(a) K-SVD (b) K-GASVD

Fig.3.1 Visualization of the color image dictionaries learned by K-SVD (a) and K-GASVD (b) methods

method captures more color information since both the relationship of spectral channels and the spatial coherence are retained better.

3.5 Experimental Analysis

To demonstrate the effectiveness and powerfulness of the proposed multivector sparse representation model for multispectral image processing, we perform both reconstruction and denoising experiments on both Columbia Multispectral Image Database[37] and Hyperspectral Image Database of Real-world Scenes[38] and compare the experimental results both quantitatively and visually.

3.5.1 Data Sets

1. Multispectral Image Datasets

The Columbia Multispectral Image Database[37] is utilized in our experiments. In this multispectral image data set, there are 32 real-world scenes consisting of various real objects and materials, each with spectral resolution 31 and spatial resolution 512×512, which includes full spectral resolution reflectance data collected from 400nm to 700nm in 10nm steps. More details can be found in http://www.cs.columbia.edu/CAVE/databases/multispectral/.

In our experiments, several arbitrary multispectral images are selected from the data set above to evaluate the performance of our K-GASVD method. Due to the limitation of the existing sparse models which can only deal with an ordinary grayscale image or color image, we randomly select 6 bands from the chosen multispectral images to form 20 pseudo color images, respectively. Fig.3.2(a)-(f), (g)-(l) and (m)-(r) show the images of band 2, 5, 7, 12, 20 and 25 belong to three multispectral images, respectively. Fig.3.2(s)-(t), (u)-(v) and (w)-(x) show the two pseudo color images in which band 2, 5, and 7 images and 12, 20, and 25 images input as the R, G and B channels, respectively.

2. Hyperspectral Image Datasets

Since hyperspectral images include per-pixel irradiance measurements in a number of narrow bands of wavelength in the visible spectrum, they provide higher spectral

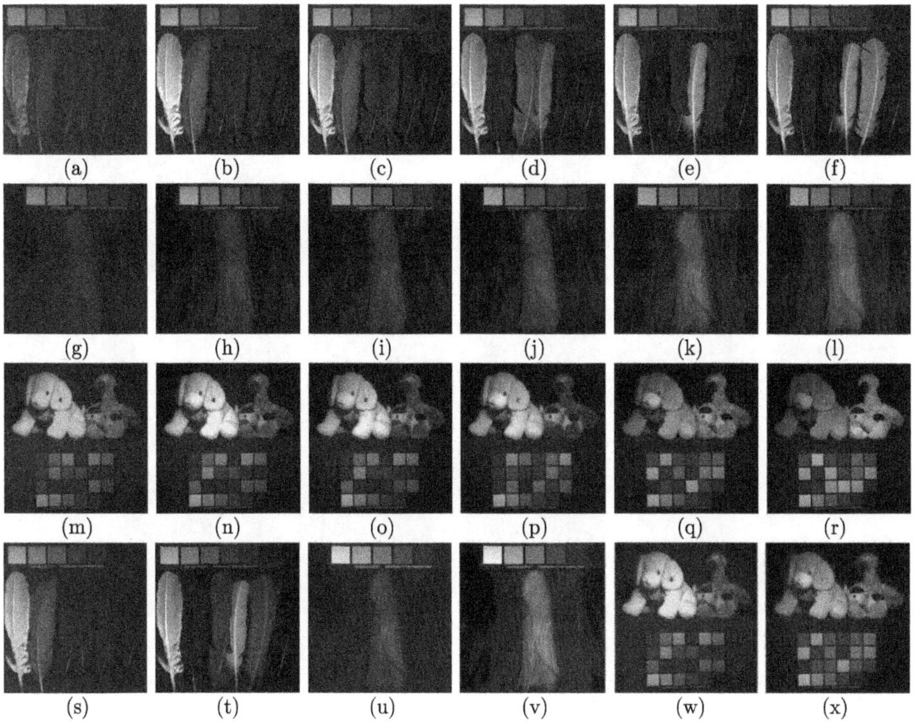

Fig.3.2 Examples of multispectral images. (a), (g), (m) band 2 images; (b), (h), (n) band 5 images; (c), (i), (o) band 7 images; (d), (j), (p) band 12 images; (e), (k), (q) band 20 images; (f), (l), (r) band 25 images; (s), (u), (w) the pseudo color images using 2, 5, and 7 band images as the RGB input; (t), (v), (x) the pseudo color images using 12, 20, and 25 band images as the RGB input

resolution and more spectral channels than multispectral images. Therefore, Hyperspectral Image Database of Real-world Scenes[38] and good performance of the proposed sparse representation model. Hyperspectral Image Database of Real-world Scenes collected fifty images under daylight illumination, both outdoors and indoors, using a commercial hyperspectral camera (Nuance FX, CRI Inc.). The camera uses an integrated liquid crystal tunable filter and is capable of acquiring a hyperspectral image by sequentially tuning the filter through a series of thirty-one narrow wavelength bands, each with approximately 10nm bandwidth and centered at steps of 10nm from 420nm to 720nm.

In our experiments, we arbitrarily select several hyperspectral images from the data set above to evaluate the performance of the proposed K-GASVD method.

Fig.3.3 shows two hyperspectral images with 6 spectral bands (band 8, 9, 15,19, 25 and 30) respectively, and some pseudo color images formed by those bands.

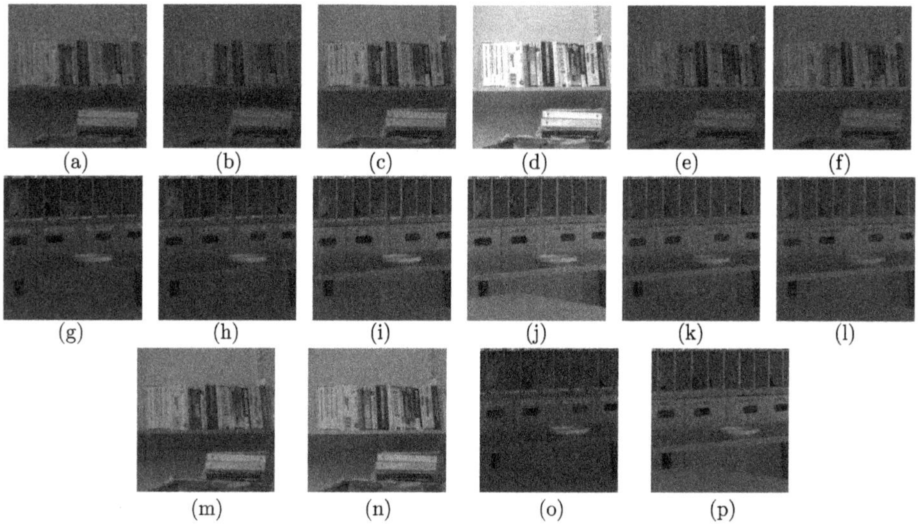

Fig.3.3 Examples of hyperspectral images. (a) and (g) band 8 images; (b) and (h) band 9 images; (c) and (i) band 15 images; (d) and (j) band 19 images; (e) and (k) band 25 images; (f) and (l) band 30 images; (m) and (o) the pseudo color images using 8, 9, and 15 band images as the RGB input; (n) and (p) the pseudo color images using 15, 19, and 30 band images as the RGB input

3.5.2 Multispectral Images Reconstruction

As described above, the K-SVD based model[11] has achieved excellent results for gray-scale images processing, but for multispectral images, it just applies K-SVD method independently onto each spectral channel, leading to unsatisfying reconstruction results.Therefore, in this section, we mainly prove the reconstruction results of the proposed K-GASVD based sparse model experimentally, and compare it with the results of K-SVD based model under the same parameter condition. Considering reasonable computational complexity and the fairness of comparison, we set the size of multispectral image patches to 8×8.

As shown in Fig.3.4, the proposed K-GASVD based sparse model significantly improves the reconstruction performance and the improvement becomes even more remarkable with respect to the increasing number of selected atoms. We further compare the number of atoms used of the two sparse models under the same PSNR. For example, it is clearly that, when the PSNR value get to nearly 34, the numbers of atoms corresponding to the K-SVD and K-GASVD models are about 450 and 150 respectively. That is to say, only approximate 1/3 number of atoms are needed in the K-GASVD model than that in K-SVD model[11] for similar reconstruction performance achievement with a reasonable sparse parameter. Clearly, the GA-based dictionary shows its capability in more colorful structures presentation, which manifests lower intra-redundancy of spectral channels of each atom in the multispectral images. Moreover, while using more dictionary, the proposed K-GASVD model reveals great performance more obviously.

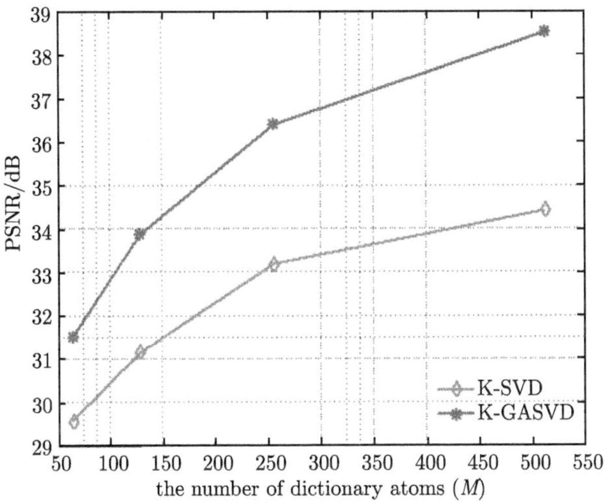

Fig.3.4 Comparison of K-GASVD and K-SVD based sparse models for multispectral images reconstruction-PSNR values vs. the number of dictionary atoms M

Then, the influence of sparse parameter is considered for both models by computing the PSNR (dB) values under different sparse parameter T, and the results are shown as two curves in Fig.3.5.

It can be clearly seen from Fig.3.5 that with the increasing number of sparse parameter T, the growth trends of the two curves are basically the same. Moreover, under the same sparse parameter T, the proposed K-GASVD model can obtain much higher PSNR values compared to the K-SVD model[11].

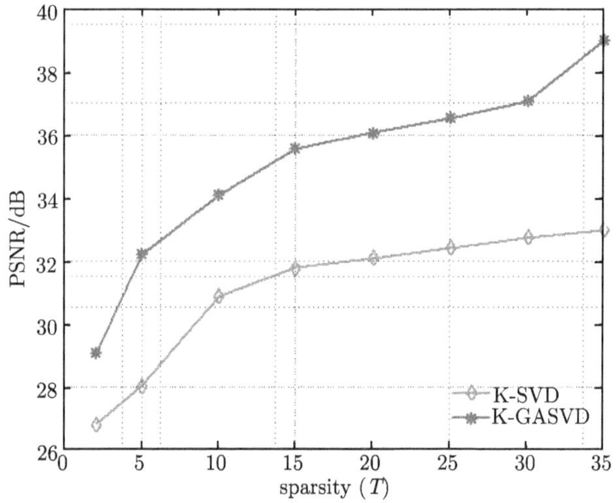

Fig.3.5 Comparison of K-GASVD and K-SVD sparse models for multispectral images reconstruction-PSNR values vs. sparse parameter T

3.5.3 Multispectral Image Denoising

Then, another application for multispectral images-denoising experiments are conducted on various multispectral images at different noise levels. First, let F_0 be a clean multispectral image represented in the GA form, denoted as a column vector of length N. Then the noisy version is

$$Y = F_0 + w \tag{3.25}$$

where $F_0 = \{f_{0i} \in (\mathbb{G}_n)^N, 1 \leqslant i \leqslant K\} \in (\mathbb{G}_n)^{N\times K}$ and ω is a white Gaussian noise in the GA form (zero means and deviation δ). Then we consider the image patches of size $\sqrt{N} \times \sqrt{N}$, which are overlapping extracted from the noisy multispectral images, ordered as column vectors $y \in (\mathbb{G}_n)^N$, and $Y = \{y_i, 1 \leqslant i \leqslant K\} \in (\mathbb{G}_n)^{N\times K}$. Handling the multispectral images denoising problem using a sparse decomposition technique per overlapped patch is then converted to solve the following minimization problem:

$$\left\{\hat{D}, \hat{a}_{kl}, \hat{F}\right\} = \min_{\hat{D},\hat{a}_{kl},\hat{F}_0} \gamma \left\|Y - F\right\|_2^2 + \sum_{k,l} \eta_{kl} \left\|a_{kl}\right\|_0 + \sum_{k,l} \left\|Da_{kl} - J_{kl}Y\right\|_2^2 \tag{3.26}$$

where $\hat{F}$ and the GA dictionary $\hat{D} \in (\mathbb{G}_n)^{N\times M}$ with M atoms denote estimators of F_0 and the optimal dictionary, respectively. Here, the optimal dictionary makes representation of recovered image patches to be sparsest. The indices $[k,l]$ indicates the location of the patch in the image. The vectors are the sparse representations for the $[k,l]$-th patch in $\hat{F}$ using the dictionary $\hat{D}$. The operator J_{kl} extracts the patch of coordinates $[k,l]$ from the image.

Clearly, the first term in Eq.(3.26) indicates the probability that requires a proximity between $\hat{F}$ and Y. The image prior are exactly posed by both the second and the third terms, in which the sparsest representation and the consistency of the decomposition are provided and ensured by the second and third terms respectively. This regularization term provides an assumption that good-characterized natural images are to exhibit the sparse representation for every patch in the image over the learned dictionary $\hat{D}$.

Considering reasonable computational complexity and the fairness of comparison, we set, the number of iterations $J = 30$, the size of color image patches to 8×8 and the number of both dictionary atoms $M = 256$.

Firstly, we validate the results achieved by applying the monochromatic model based on K-SVD method[11], the concatenation model based on improved K-SVD method[16] and our multivector model based on K-GASVD method, on the testing multispectral images shown in Fig.3.2 and hyperspectral images shown in Fig.3.3 with the noise deviation $\delta = 30$, in order to enable a fair comparison. Fig.3.6 and Fig.3.7 show the comparison of the K-SVD, the improved K-SVD and our proposed K-GASVD models in terms of denoising results. Notice that, for the multispectral and hyperspectral images, there is no existing sparse model, the general method is to regard each single-band image of the multispectral or hyperspectral image as a gray-scale image to process them respectively, such as the K-SVD method, or concatenate all the spectral channels into a single vector using a shared sparse coefficient, such as the improved KSVD method. As shown in Fig.3.6 and Fig.3.7 the K-SVD based model is performed in a monochromatic way, which ignores the correlation among these spectral channels and loses the inherent spectral structures, which introduces a hue bias and leads to a lack of details. The improved K-SVD based sparse model reduces some hue bias, but still loses channel interrelationship and produces some false colors and artifacts.

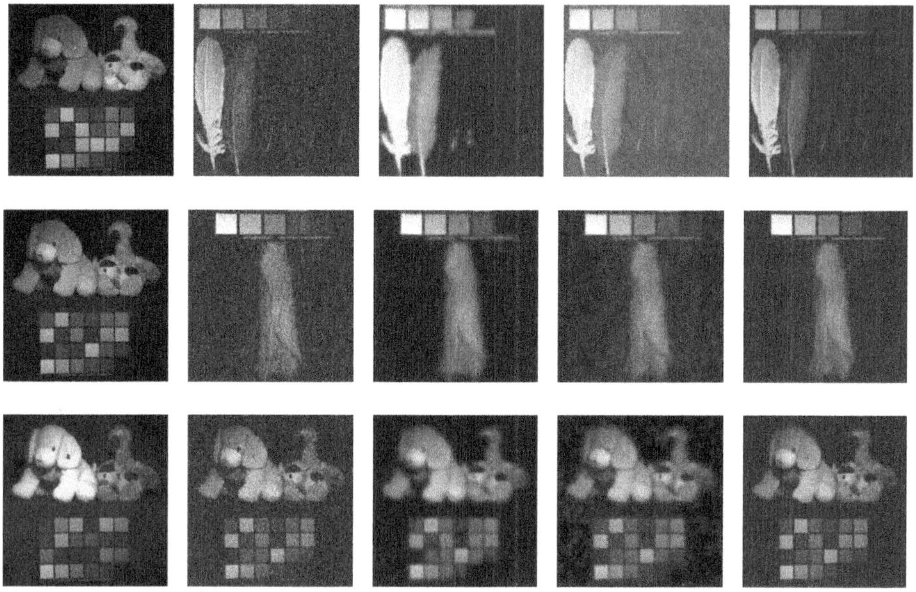

(a) Original images (b) Noisy images ($\delta = 30$) (c) K-SVD[11] (d) Improved K-SVD[16] (e) K-GASVD

Fig.3.6 Denoising results of multispectral images

(a) Oiginal images (b) Noisy images ($\delta = 30$) (c) K-SVD[11] (d) Improved K-SVD[16] (e) K-GASVD

Fig.3.7 Denoising results of hyperspectral images

In the high-frequency texture areas, our K-GASVD sparse model that regards a multispectral or hyperspectral image as a GA multivector, has a clear advantage over the K-SVD and the improved K-SVD methods that either represent each single-band image as a scalar or concatenate all the spectral channels into a single channel.

We now turn to study the influence of different noise levels. To evaluate the superiority of our K-GASVD method objectively, the denoising results represented by PSNR and SSIM, which are achieved by applying our proposed method, the improved K-SVD method and the K-SVD method over several noise levels on the same testing multispectral and hyperspectral images are collected in Table 3.3 and Table 3.4. The noise deviation δ are set to be 15, 20, 25 and 30. It is shown

Table 3.3 PSNR values (dB) obtained by different denoising methods at each noisy level based on multispectral image datasets[37]

δ	methods	balloon	pepper	feather	superball	flower	bead
15	K-SVD[11]	36.61	35.77	35.39	36.27	36.66	35.22
	improved K-SVD[16]	36.89	36.08	36.42	36.90	37.01	36.05
	K-GASVD	39.51	37.37	38.69	38.49	39.38	38.32
20	K-SVD[11]	34.31	34.42	32.62	34.51	34.49	33.17
	improved K-SVD[16]	34.89	34.67	32.98	35.01	35.32	34.03
	K-GASVD	38.70	36.07	37.11	37.33	37.71	36.71
25	K-SVD[11]	32.60	31.63	31.34	30.82	31.36	31.09
	improved K-SVD[16]	32.93	31.95	32.02	31.23	31.94	31.64
	K-GASVD	34.57	32.35	33.29	33.00	35.45	34.43
30	K-SVD[11]	29.88	28.59	28.58	28.51	29.25	28.87
	improved K-SVD[16]	29.92	28.89	28.87	28.82	29.67	29.34
	K-GASVD	30.96	30.05	30.55	30.28	30.65	30.56

Table 3.4 SSIM values obtained by different denoising methods at each noisy level based on hyperspectral image datasets[38]

δ	methods	img3	img4	imgd0	imgd5	imgg8	imgg9
15	K-SVD[11]	0.8738	0.8743	0.8404	0.8621	0.8878	0.8326
	improved K-SVD[16]	0.8757	0.8862	0.8496	0.8636	0.8941	0.8902
	K-GASVD	0.9073	0.9073	0.8869	0.8949	0.9105	0.9102
20	K-SVD[11]	0.8649	0.8649	0.8343	0.8532	0.8419	0.7859
	improved K-SVD[16]	0.8670	0.8670	0.8394	0.8546	0.8506	0.8397
	K-GASVD	0.8967	0.8967	0.8671	0.8833	0.8913	0.8697
25	K-SVD[11]	0.8249	0.8249	0.7809	0.8202	0.7868	0.7529
	improved K-SVD[16]	0.8306	0.8306	0.7775	0.8242	0.8199	0.7889
	K-GASVD	0.8715	0.8715	0.8827	0.8504	0.8370	0.8689
30	K-SVD[11]	0.7918	0.7918	0.7268	0.7949	0.7365	0.7239
	improved K-SVD[16]	0.7989	0.7989	0.7315	0.7979	0.7952	0.7408
	K-GASVD	0.8605	0.8605	0.8585	0.8382	0.8229	0.8408

in Table 3.3 and Table 3.4, the improved K-SVD method shows better denoising results compared to the K-SVD method, for it partly reserves the inherent spectral structure with the use of concatenated channel. However, it still produces some hue bias and artifacts. Our proposed method outperforms the results of the K-SVD and the improved K-SVD methods, better results are achieved with small values of δ and vice versa. In fact, this is expected because relatively "clean" images have a greater impact on the outcome, on the contrary, very noisy images have a weak influence on the results, if any. It is mainly because our proposed K-GASVD method attains the inherent spectral structure of multispectral images to reserve more details and avoid the artifacts.

3.6 Conclusions

This work has put forward a new sparse representation model based on geometric algebra (GA) for multispectral images processing, resulting in state-of-the-art performance and exceeding recent leading alternatives. The proposed sparse model treats a multispectral image as a GA multivector, totally preserves the relationships of the spectral channels to avoid the loss of inherent spectral structures. Besides, we provide a GA-based dictionary learning method, called K-GASVD method, which can capture the important spectral spacerelated information while capturing the important spectral structure information, and replace the sparse dictionary of several channels with a GA dictionary. The reconstruction and denoising experiments demonstrate the superiority of the proposed model and show its potential in multispectral images processing and analysis.

Appendix A

A. Derivation of the Generalized Form of Sparse Representation Model Based on GA

Then, the generalized form of representation model for multispectral images can be obtained as follows

$$\begin{aligned}&[0 \quad \cdots \quad E_i(f) \quad \cdots \quad E_{ij}(f) \quad \cdots \quad E_{1\cdots n}(f)]\\&=[E_0(D) \quad \cdots \quad E_i(D) \quad \cdots \quad E_{ij}(D) \quad \cdots \quad E_{1\cdots n}(D)]\,\tilde{a}\end{aligned} \tag{3.27}$$

According to the relationship of a multispectral image $F \in (\mathbb{G}_n)^{N\times K}$, the GA dictionary $D \in (\mathbb{G}_n)^{N\times M}$ and the sparse coefficient vector $a \in (\mathbb{G}_n)^M$ shown in Eq.(3.28).

Therefore, we can obtain the generalized sparse coefficient matrix $\tilde{a} \in \mathbb{R}^{2^n M\times 2^n N}$ shown in Eq.(3.29).

$$
\begin{aligned}
f =Da \Leftrightarrow\ & 0 + \sum_{1\leqslant i\leqslant n} E_i(f)e_i + \sum_{1\leqslant i<j\leqslant n} E_{ij}(f)e_{ij} + \cdots + E_{1\cdots n}(f)e_{1\cdots n} \\
= & \Bigg(E_0(D)E_0(a) + \sum_{1\leqslant i\leqslant n} E_i(D)E_i(a) - \sum_{1\leqslant i<j\leqslant n} E_{ij}(D)E_{ij}(a) \\
& + \cdots + (-1)^{n(n-1)/2} E_{1\cdots n}(D)E_{1\cdots n}(a) \Bigg) e_0 \\
& + \cdots + \left(\begin{array}{c} E_t(D)E_0(a) + \dfrac{i-t}{|i-t|} \displaystyle\sum_{1\leqslant i\leqslant n} E_{ti}(D)E_i(a) \\ +E_0(D)E_t(a) + \dfrac{t-i}{|t-i|} \displaystyle\sum_{1\leqslant i\leqslant n} E_i(D)E_{it}(a) \\ + \cdots + (-1)^{(t-1)+(n(n-1)/2)} E_{1\cdots(t-1)(t+1)\cdots n}(D)E_{1\cdots n}(a) \end{array} \right) e_t \\
& + \cdots + \left(\begin{array}{c} E_{ts}(D)E_0(a) + \dfrac{(i-t)(i-s)}{|i-t|\,|i-s|} \displaystyle\sum_{1\leqslant i\leqslant n} E_{tsi}(D)E_i(a) + E_t(D)E_s(a) \\ -E_s(D)E_t(a) + \dfrac{(i-t)(i-s)(j-t)(j-s)}{|i-t|\,|i-s|\,|j-s|\,|j-t|} \displaystyle\sum_{1\leqslant i<j\leqslant n} E_{tsij}(D)E_{js}(a) \\ + \dfrac{(j-t)(s-j)}{|j-t|\,|s-j|} \displaystyle\sum_{1\leqslant j\leqslant n} E_{tj}(D)E_{js}(a) + \cdots + (-1)^{(t-1)+(s-2)+(n(n-1)/2)} \\ E_{1\cdots(t-1)(t+1)\cdots(s-1)(s+1)\cdots n}(D)E_{1\cdots n}(a) \end{array} \right) e_{ts} \\
& + \cdots + \left(\begin{array}{c} E_{1\cdots n}(D)E_0(a) + (-1)^{(n-i)} \displaystyle\sum_{1\leqslant i\leqslant n} E_{1\cdots(i-1)(i+1)\cdots n}(D) \\ E_i(a) + (-1)^{(2n-i-j+1)} \displaystyle\sum_{1\leqslant i<j\leqslant n} E_{1\cdots(i-1)(i+1)\cdots(j-1)(j+1)\cdots n}(D) \\ E_{ij}(a) + E_0(D)E_{1\cdots n}(a) \end{array} \right) e_{1\cdots n}
\end{aligned}
\tag{3.28}
$$

where $e_0 = 1; t = 1,\cdots,n; i \neq t, i \neq s, j \neq t, j \neq s; j > i, s > t; E(D)_{ij} = E(D)_{ji}; E(Da)_{ij} = E(Da)_{ji}$.

$$\tilde{a}=\begin{bmatrix} E_0(a) & \cdots & E_t(a) & \cdots & -E_{ts}(a) & \cdots & (-1)^{n(n-1)/2}E_{1\cdots n}(a) \\ \vdots & & \vdots & & \vdots & & \vdots \\ E_t(a) & \cdots & E_0(a) & \cdots & -E_s(a) & \cdots & (-1)^{(t-1)+(n(n-1)/2)}E_{1\cdots n}(a) \\ \vdots & & \vdots & & \vdots & & \vdots \\ E_{ts}(a) & \cdots & -E_s(a) & \cdots & E_0(a) & \cdots & (-1)^{(t-1)+(s-2)+(n(n-1)/2)}E_x(a) \\ \vdots & & \vdots & & \vdots & & \vdots \\ E_{1\cdots n}(a) & \cdots & (-1)^{(n-t)}E_y(a) & \cdots & (-1)^{(2n-t-s+1)}E_x(a) & \cdots & E_0(a) \end{bmatrix} \tag{3.29}$$

where $x=1\cdots(t-1)(t+1)\cdots(s-1)(s+1)\cdots n; y=1\cdots(t-1)(t+1)\cdots n; t=1\cdots n; i\neq t, t\neq s, j\neq t, j\neq s; j>i, s>t; E_{ij}=E_{ji}$.

References

[1] Said B, Hadjidj R, Melkemi K E, et al. Multispectral image denoising with optimized vector non-local mean filter. Digital Signal Processing, 2016, 58(C): 115-126.

[2] Tzoumas S, Rosenthal A, Lutzweiler C, et al. Spatio spectral denoising framework for multispectral optoacoustic imaging based on sparse signal representation. Medical Physics, 2014, 41(11): 113301.

[3] Xie Q, Zhao Q, Meng D, et al. Multispectral images denoising by intrinsic tensor sparsity regularization. Computer Vision & Pattern Recognition. 2016: 1692-1700.

[4] Valsesia D, Magli E. A novel rate control algorithm for onboard predictive coding of multispectral and hyperspectral images. IEEE Transactions on Geoscience and Remote Sensing, 2014, 52(10): 6341-6355.

[5] Valsesia P, Jeyakumar S, Nithiyanandam N. Hyperspectral image compression based on non-iterative matrix factorization. 2013 IEEE International Conference on Computational Intelligence and Computing Research (ICCIC). Piscataway: IEEE, 2013.

[6] Kawakami R, Matsushita Y, Wright J, et al. High-resolution hyperspectral imaging via matrix factorization. Computer Vision and Pattern Recognition, 2011, 42: 2329-2336.

[7] Chen A. The inpainting of hyperspectral images: A survey and adaptation to hyperspectral data. Image and Signal Processing for Remote Sensing XVIII, 2011, 8537: 85371K-85371K-8.

[8] Nguyen H V, Banerjee A, Chellappa R. Tracking via object reflectance using a hyperspectral video camera. IEEE Computer Society Conference on Computer Vision and Pattern Recognition - Workshops IEEE, 2011: 44-51.

[9] Aharon M, Elad M, Bruckstein A M. The K-SVD: An algorithm for designing of overcomplete dictionaries for sparse representations. IEEE Trans. Image Process, 2006, 54(11): 4311-4322.

[10] Wright J, Ma Y, Mairal J, et al. Sparse representation for computer vision and pattern recognition. Proc. IEEE, 2010, 98(6): 1031-1044.

[11] Elad M, Aharon M. Image denoising via sparse and redundant representations over learned dictionaries. IEEE Trans. Image Process, 2006, 15(12): 68-80.

[12] Guleryuz G. Nonlinear approximation based image recovery using adaptive sparse reconstructions and iterated denoising-Part I: Theory. IEEE Trans. Image Process, 2006, 15(3): 539-554.

[13] Wang Z C, Jiang R J, Jiang X F, et al. Learning sparse representations by K-SVD for facial expression classification. 2015 4th International Conference on Computer Science and Network Technology (ICCSNT), 2015: 772-775.

[14] Chen F, Yu H M, Hu R. Shape sparse representation for joint object classification and segmentation. IEEE Trans. Image Process, 2013, 22(3): 992-1004.

[15] Agarwal S, Roth D. Learning a sparse representation for object detection. European Conf. on Computer Vision (ECCV), 2006: 97-101.

[16] Mairal J, Elad M, Sapiro G. Sparse representation for color image restoration. IEEE Trans. Image Process, 2007, 17(1): 53-69.

[17] Xu Y, Yu L C, Xu H T, et al. Vector sparse representation of color image using quaternion matrix analysis. IEEE Trans. Image Process, 2015, 24(4): 1315-1328.

[18] Wang R, Wu Y J, Shen M M, et al. Sparse representation for color image based on geometric algebra. 2018 IEEE International Conference on Multimedia and Expo (ICME), 2018: 1-6.

[19] Shen M M, Wang R, Cao W M. Joint sparse representation model for multi-channel image based on reduced geometric algebra. IEEE Access Image Process, 2018, DOI: 10.1109/ACCESS.2018.2819691.

[20] Shen M M, Wang R. New Singular Value Decomposition Algorithm for Octonion Signals. 24th International Conference on Pattern Recognition (ICPR), 2018: 3233-3237.

[21] Yu L C, Xu Y, Xu H T, et al. Quaternion-based sparse representation of color image. Proc. IEEE Int. Conf. Multimedia Expo (ICME), 2013: 1-7.

[22] Barthelemy Q, Larue A, Mars J I. Color sparse representations for image processing: Review, models, and prospects. IEEE Trans. Image Process, 2015, 24(11): 3978-3989.

[23] Franchini S, Gentile A, Sorbello F, et al. Conformal ALU: A conformal geometric algebra coprocessor for medical image processing. IEEE Transactions on Computers, 2015, 64(4): 955-970.

[24] Su H, Bo Z. Conformal Geometric Algebra Based Band Selection and Classification for Hyper Spectral Imagery. 2016 8th Workshop on Hyperspectral Image and Signal Processing: Evolution in Remote Sensing (WHISPERS), Los Angeles, CA, 2016: 1-4.

[25] Pham M T, Yoshikawa T, Furuhashi T, et al. Robust feature extractions from geometric data using geometric algebra. IEEE International Conference on Systems Man and Cybernetics, 2009: 529-533.

[26] Ebling J, Scheuermann G. Clifford Fourier transform on vector fields. IEEE Transactions on Visualization & Computer Graphics, 2005, 11(4): 469-479.

[27] Wang R, Zhou Y X, Jin Y L, et al. Sparse fast Clifford Fourier transform. Frontiers of Information Technology & Electronic Engineering, 2017, 18(8): 1131-1141.

[28] Li Y, Liu W, Li X, et al. GA-SIFT: A new scale invariant feature transform for multispectral image using geometric algebra. Information Sciences, 2014: 559-572.

[29] Huang B, Song H, Cui H, et al. Spatial and spectral image fusion using sparse matrix factorization. IEEE Transactions on Geoscience and Remote Sensing, 2014, 52(3): 1693-1704.

[30] Lanaras C, Baltsavias E, Schindler K. Advances in hyperspectral and multispectral image fusion and spectral unmixing. International Archives of the Photogrammetry Remote Sensing, 2015, 3(3): 451-458.

[31] Dian R, Fang L, Li S. Hyperspectral Image Super-resolution via Non-local Sparse Tensor Factorization. 2017 IEEE Conference on Computer Vision and Pattern Recognition (CVPR), 2017: 3862-3871.

[32] Zhang K, Wang M, Yang S, et al. Spatial-spectral-graph-regularized low-rank tensor decomposition for multispectral and hyperspectral image fusion. IEEE Journal of Selected Topics in Applied Earth Observations & Remote Sensing, 2018: 1-11.

[33] Shih Y T, Chien C S, Chuang C Y. An adaptive parameterized block-based singular value decomposition for image de-noising and compression. Applied Mathematics & Computation, 2012, 218(8): 10370-10385.

[34] Huang J, Xie J, Li F, et al. Compound de-noising method based on discrete stationary

wavelet transform and optimized SVD. Piezoelectrics & Acoustooptics, 2013, 35(3): 448-451.

[35] Chen L, Zhang X, Shen Y J, et al. Fault diagnosis of rolling bearings based on SVD denoising and blind signals separation. Journal of Vibration & Shock, 2012, 31(23): 185-190.

[36] Pati Y C, Rezaiifar R, Krishnaprasad P S. Orthogonal matching pursuit: Recursive function approximation with applications to wavelet decomposition. Proc. Conf. Rec. 27th Asilomar Conf. Signals, Syst. Comput., 1993: 40-44.

[37] Yasuma F, Mitsunaga T, Iso D, et al. Generalized assorted pixel camera: Postcapture control of resolution, dynamic range, and spectrum. IEEE Trans. Image Processing, 2010, 19(9): 2241-2253.

[38] Chakrabati A, Zickler T. Statistics of Real-World Hyperspectral Images. Proceedings of the IEEE Conference on Computer Vision and Pattern Recognition (CVPR), 2011.

Chapter 4

GA-SURF: A New Speeded-up Robust Feature Extraction Algorithm for Multispectral Images Based on Geometric Algebra

4.1 Introduction

Multispectral images are widely used in more and more fields[1−5], including computer vision, medicine, biology, astronomy, remote sensing and digital photography. Multispectral images can provide more information than single channel images. Accordingly, multispectral images processing has become a new research hotspots[6,7], and feature analysis plays an important role in many multispectral image applications[1,3,5−9], including object recognition, image registration and classification.

Scale invariant feature transform (SIFT)[10,11] is the most popular method for the local invariant features extraction. C-SIFT[12] is widely used color image local invariant feature extraction algorithm, which introduces a novel composite local feature consisting of color information and local texture. It has shown that C-SIFT has better performance than SIFT. Since then, many color image feature extraction algorithms have been developed, such as Hue-SIFT[13], HSV-SIFT[14], opponent SIFT[14], rgSIFT[14], RGB-SIFT[14], CIC-SIFT[15], and so on.

However, such algorithms cannot be directly applied to multispectral images with more than three bands. To solve this problem, Li et al. introduced geometric algebra (GA)[16−20] to propose a new representation of multispectral image including spatial and spectral information, and develop a new scale invariant feature transform (GA-SIFT) method[21]. Taking advantage of the GA theory, a multispectral image can be regarded as an embedding of a manifold in a higher dimensional spectral-spatial space. Both of the spectral and spatial information of the multispectral image can be retained. The experiments show that GA-SIFT outperforms some other SIFT-based feature extraction methods, such as standard SIFT, Hue-SIFT, C-SIFT.

Though GA-SIFT is powerful for multispectral image feature analysis, it has the drawback of large computation burdens. As is known, SURF[22,23] is a computationally-efficient replacement to SIFT, which approximates or even outperforms some previous SIFT algorithms with respect to distinctiveness and robustness in extracting and matching interest points, and it can be computed much faster. Although extended SURF algorithms have been proposed including SURF-DAISY[24], Color-SURF[25] and SSURF[26], however, they cannot be applied directly for multispectral images.

Because the existing SURF algorithms are mostly developed for gray or 3-channel color images, they cannot extract features efficiently from multispectral images. In this chapter, we pay attention to this issue and aim to develop both a detector and descriptor, which in comparison to the state-of-the-art algorithms are faster to compute, while not sacrificing performance. Based on the GA representation of the multispectral images in GA-SIFT[21], we propose a novel SURF algorithm called GA-SURF for multispectral images. GA-SURF incorporates GA theory[16−20] into traditional SURF framework. We first propose a new Hessian matrix in GA to locate the interest points and utilize integral images to simplify its calculation with less accuracy loss. Then descriptors are generated from neighboring pixels. Experimental results show that GA-SURF is robust and fast for feature computation and matching.

4.2 Related Work

4.2.1 SURF Algorithm

The SURF algorithm[22,23] proposed by Herbert Bay et al. in 2006 can effectively extract feature information. The process of the SURF algorithm can be described as follows.

Stage 1 The Hessian matrix is defined as

$$H\left(L\left(x,y\right),\sigma\right)=\begin{bmatrix}\dfrac{\partial^2 L}{\partial x^2} & \dfrac{\partial^2 L}{\partial x\partial y}\\ \dfrac{\partial^2 L}{\partial x\partial y} & \dfrac{\partial^2 L}{\partial y^2}\end{bmatrix} \tag{4.1}$$

where, $L(x,y)$is the convolution of the Gaussian second order derivative with the input signal $f(x,y)$, and the variance of the Gaussian filter $g(x,y,\sigma)$ is σ^2.

In order to simplify the computation of Hessian matrix determinants, the Gaussian second order derivatives are transformed to box filters and the approximation of Gaussian second order derivatives D_{xx}, D_{yy}, D_{xy} can be obtained. Approximate Hessian matrix determinant is computed as

$$\det\left(H_{\text{approx}}\right)=D_{xx}D_{yy}-\left(\omega D_{xy}\right)^2 \tag{4.2}$$

where, ω is the weighted coefficient.

Stage 2 The image pyramid is constructed using the scale-varied box filters. In SURF, 9×9 size filter is used as the starting filter, and then the filter size L can be calculated by

$$L=3\times\left(2^{\text{Octave}}\times\text{interval}+1\right) \tag{4.3}$$

where, the value of Octave and interval starts from 1.

While using the integral images with size L to approximate 2D Gaussian kernel filtering, the corresponding 2D Gaussian kernel parameters are

$$\sigma=\sigma_0\times\frac{L}{9} \tag{4.4}$$

Stage 3 In order to locate interest points in image pyramid, non-maximal suppression is used in a $3\times 3\times 3$ neighborhood. Every pixel should be compared with the remaining 26 pixels, which are eight pixels of its own scale layer and two nine-pixels on the above and below scale layers. If the pixel value is greater than the surrounding pixels, the pixel can be determined as an interest point.

Stage 4 Finally, the SURF descriptor is generated. We first obtain the dominant orientation of interest points and then calculate SURF descriptor by using Haar wavelet transform.

4.2.2 The Basics of Geometric Algebra

Geometric algebra (GA) provides an efficient framework to conduct the computations without using the coordinate information[16-20]. It is easy to represent complex operations byusing k-vectors. There are all geometric operators in GA. It has widely been utilized in physics and computer vision.

$\mathbb{G}_n$ is defined as n-dimensional geometric algebra space. In $\mathbb{G}_n$, there are n orthogonal basis which can be represented as $e_1, e_2, \cdots, e_n$. If $x \in \mathbb{G}_n$, x can be represented by

$$x = \sum_{i=1}^{n} x_i e_i \tag{4.5}$$

where $x_1, x_2, \cdots, x_n \in \mathbb{R}$.

Obviously, for $n > 1$, the multiplication is not commutative in $\mathbb{G}_n$. The basis $e_1, e_2, \cdots, e_n$ has many operations, as follows.

$$e_i \wedge e_j = -e_j \wedge e_i, \quad i \neq j, \quad i, j = 1, 2, \cdots, n \tag{4.6}$$

$$e_i^2 = 1, \quad i, j = 1, 2, \cdots, n \tag{4.7}$$

$$e_i \cdot e_j = \frac{1}{2}(e_i e_j + e_j e_i), \quad i, j = 1, 2, \cdots, n \tag{4.8}$$

$$e_i e_j = e_i \cdot e_j + e_i \wedge e_j, \quad i, j = 1, 2, \cdots, n \tag{4.9}$$

where $\cdot$ is the inner product symbol, $\wedge$ is the outer product symbol. e_i, e_j represents the product of e_i and e_j. It is equal to the inner product plus the outer product.

GA provides an efficient framework to conduct computations independent of the coordinate system. In the framework of GA, all geometric objects are represented as multivectors. Vector calculations can be implemented by algebraic calculations without coordinate information. GA makes it easy to compute in high dimensional space.

4.2.3 GA-SIFT Algorithm

Since existing SIFT algorithms are mostly developed for gray or 3-channel color images, they are not suitable for multispectral images. In order to solve the problem, GA-SIFT[21] take advantage of multi-channel information in multispectral images based on GA theory. The processes of GA-SIFT are as follows.

Step 1 Obtain the chromaticity part $L_c(x, y, \Sigma)$ of an input multispectral image $f(x, y)$ the 2-side convolution and φ function.

Step 2 Employ the scale-varied Gaussian filters in GA to obtain the scale space of the multispectral image L_c.

$$L_c = \{L_c(x, y, k\Sigma)|k = 1, 2, \cdots, m\} \tag{4.10}$$

where $L_c(x, y, k\Sigma)$ can be calculated as

$$L_c(x, y, k\Sigma) = \varphi(G_{n+2}(x, y, k\Sigma) \otimes f(x, y) \otimes G_{n+2}(x, y, k\Sigma)) \tag{4.11}$$

Then, we can get GA-DoG images by using L_c.

Step 3 Find local minima/maxima of the GA-DoG images as key points. The details can be found in Ref. [21].

Step 4 Generate the GA-SIFT descriptor for each detected interest points using the magnitude (Eq. (4.14)) and orientation (Eq. (4.15)) in GA.

$$\Delta x = L_c(x + 1, y, \Sigma) - L_c(x - 1, y, \Sigma) \tag{4.12}$$

$$\Delta y = L_c(x, y + 1, \Sigma) - L_c(x, y - 1, \Sigma) \tag{4.13}$$

$$m(x, y) = \sqrt{(|\Delta x|)^2 + (|\Delta y|)^2} \tag{4.14}$$

$$\theta(x, y) = \arccos\left(\frac{\Delta x \cdot \Delta y}{|\Delta x||\Delta y|}\right) \tag{4.15}$$

4.3 The Proposed GA-SURF Algorithm

4.3.1 The Construction of the Hessian Matrix

According to Ref. [17, 21], a multispectral image can be represented as Eq. (4.16)

$$f(x, y) = \sum_{i=1}^{n} f_i(x, y)\, e_i \tag{4.16}$$

where $f(x, y)$ represents a multispectral image. $f_i(x, y)$ is the data set of the i-th band of $f(x, y)$, (x, y) denotes the 2D coordinates, and n denotes the number of bands in a multispectral image. According to Ref. [10, 11], the Gaussian filter in G_n can be represented as Eq. (4.17)

$$G_{n+2}\left(x, y, \sum(i)\right) = \sum_{i=1}^{n}\left(\frac{1}{2\pi(\Sigma(i))^2} \mathrm{e}^{-\frac{x^2+y^2}{2(\Sigma(i))^2}}\right) e_i \tag{4.17}$$

where in $n+2$, n represents the number of bands in multispectral image. 2 represents the number of coordinates in space. $\Sigma(i) = \sigma_i$, $0 \leqslant i \leqslant n$ σ_i is the standard deviation of the i-th band in the multispectral image. In fact, the Gaussian convolution kernel is used in the convolution of the multispectral image. We can get the Gaussian convolution kernel from Eq. (4.17) as

$$G_{n+2} = \begin{pmatrix} g_{11} & g_{12} & \cdots & g_{1N} \\ g_{21} & g_{22} & \cdots & g_{2N} \\ \vdots & \vdots & \ddots & \vdots \\ g_{M1} & g_{M2} & \cdots & g_{MN} \end{pmatrix} \tag{4.18}$$

where $g_{ij} = \sum\limits_{k=1}^{n} g_{ijk} e_k$ and $g_{ijk} \in \mathbb{R}$.

We introduce the method of the 2-side convolution to reflect the relationship of spectral bands in a multispectral image. $L(x, y)$ can be represented as

$$\begin{aligned} L(x,y) &= G_{n+2} \otimes f \otimes G_{n+2} \\ &= \frac{1}{M^2N^2} \sum_{l=0}^{M-1} \sum_{s=0}^{N-1} \sum_{p=0}^{M-1} \sum_{q=0}^{N-1} \left(\sum_{i=1}^{n} \sum_{j=1}^{n} g_i f_i g_j' e_j \right. \\ &\quad \left. + \sum_{i=1}^{n} \sum_{j=1}^{n} \sum_{k=1}^{n} g_i f_j g_k' \left(e_i \Lambda e_j \Lambda e_k \right) \right) \end{aligned} \tag{4.19}$$

There are two multivectors in the result of the Eq. (4.19), the second multivector is not in $\mathbb{G}_n$. The second multivector cannot be used in the image processing simply. So, we introduce a function φ to get the first part of the result of the Eq. (4.19)[21]. The definition of the function φ is as Eq. (4.20).

$$\begin{aligned} &\varphi\left(G_{n+2} \otimes f \otimes G_{n+2}\right) \\ &= \sum_{j=1}^{n} \left(\frac{1}{M^2N^2} \sum_{l=0}^{M-1} \sum_{s=0}^{N-1} \sum_{p=0}^{M-1} \sum_{q=0}^{N-1} \sum_{i=1}^{n} g_i f_i g_j' e_j \right) \end{aligned} \tag{4.20}$$

Definition 1 The Hessian matrix in GA is defined as Eq. (4.21).

$$H(x,y,\sigma) = \begin{bmatrix} \varphi\left(L_{xx}(x,y,\sigma)\right) & \varphi\left(L_{xy}(x,y,\sigma)\right) \\ \varphi\left(L_{xy}(x,y,\sigma)\right) & \varphi\left(L_{yy}(x,y,\sigma)\right) \end{bmatrix} \tag{4.21}$$

where $L_{xx}(x,y,\sigma)$, $L_{yy}(x,y,\sigma)$ and $L_{xy}(x,y,\sigma)$ are the second-order derivatives of $L(x, y)$.

Definition 2 The determinant of the Hessian matrix in GA is defined as Eq. (4.22).

$$\det(H) = |\varphi(L_{xx}) \cdot \varphi(L_{yy}) - \varphi(L_{xy}) \cdot \varphi(L_{xy})| \tag{4.22}$$

According to SURF algorithm, we put forward a new method to obtain the determinant of the Hessian matrix for improving the operation speed. We use box filters to approximate the second order derivatives of the first or second Gaussian filter. The determinant of the Hessian matrix in GA can be approximated as Eq. (4.23).

$$\det(H) = |D_{xx} \cdot D_{yy} - (\omega)^2 D_{xy} \cdot D_{xy}| \tag{4.23}$$

where ω is the weighted coefficient. ω is usually equal to 0.9. D_{xx}, D_{yy} and D_{xy} which are the approximate of Gaussian second order derivatives can be obtained by the box filters.

4.3.2 Detection and Descriptor of Interest Points in a Multispectral Image

We construct image pyramid in GA to detect interest points in a multispectral image by GA-SURF algorithm. Following the procedures of SURF algorithm, image pyramid is constructed using scale-varied box filters in GA. And the filter size can be calculated as

$$\text{Filter Size} = 3 \times \left(2^{\text{Octave}} \times \text{interval} + 1\right) \tag{4.24}$$

and the scale s can be obtained as Eq. (4.25).

$$s = 2^{\text{Octave}-1} \times \text{interval} \tag{4.25}$$

In GA-SURF, the non-maximum suppression in a $3 \times 3 \times 3$ neighborhood is applied to localize interest points in the image pyramids.

Then, in a circular neighborhood of radius $6s$ (s represents the image scale at which the interest point is detected) around the interest point, we can get the response value of the Haar wavelet in GA by using a sliding orientation window covering an angle of $\pi/3$, and the response value is the sum of $\mathrm{d}x$ and $\mathrm{d}y$. And we can get the sum m_w and corresponding orientation of the interest point as follows.

$$m_{w_x} = \sum_{i=1}^{n} \left(\sum_{w} \mathrm{d}x\right) e_i \tag{4.26}$$

$$m_{w_y} = \sum_{i=1}^{n} \left(\sum_{w} \mathrm{d}y \right) e_i \tag{4.27}$$

$$m_w = | m_{w_x} + m_{w_y} | \tag{4.28}$$

$$\theta_w = \arctan \left(| m_{w_x} | / | m_{w_y} | \right) \tag{4.29}$$

where w represents the sliding orientation window. Then we can rotate the sliding orientation window to get another m_w. When the m_w is the maximum, the orientation of the sliding window is the dominant orientation of the interest point.

In the dominant orientation of the interest point, we can get the GA-SURF descriptor by using Haar wavelet transform. The descriptor is represented as Eq. (4.30).

$$V = \left\{ \left| \sum_{i=1}^{n} \left(\sum_{w} \mathrm{d}x \right) e_i \right|, \left| \sum_{i=1}^{n} \left(\sum_{w} | \mathrm{d}x | \right) e_i \right|, \left| \sum_{i=1}^{n} \left(\sum_{w} \mathrm{d}y \right) e_i \right|, \left| \sum_{i=1}^{n} \left(\sum_{w} | \mathrm{d}y | \right) e_i \right| \right\} \tag{4.30}$$

where n is the number of bands in the multispectral image, V is a four-dimensional descriptor vector. e_i is a orthogonal basis in $\mathbb{G}_n$.

4.3.3 The Implementation of GA-SURF

Fig.4.1 shows a flow chart for the GA-SURF algorithm. Meanwhile, we summarize the proposed algorithm (GA-SURF) and describe shortly the process of this algorithm as follows.

Step 1 Extract the chromaticity image $\varphi(L(x, y))$ from an input multispectral image $f(x, y)$ in $\mathbb{G}_n$ by Eq. (4.19) and Eq. (4.20).

Step 2 Use the box filter method to obtain the determinant of the Hessian matrix in GA by Eq. (4.23).

Step 3 Construct image pyramid in GA by scale-varied box filters.

Step 4 Utilize the non-maximum suppression in a $3 \times 3 \times 3$ neighborhood to localize interest points.

Step 5 Generate the descriptor (Eq.(4.24)) for each interest point by Haar wavelet transform (Eq.(4.26)~(4.29)) in GA.

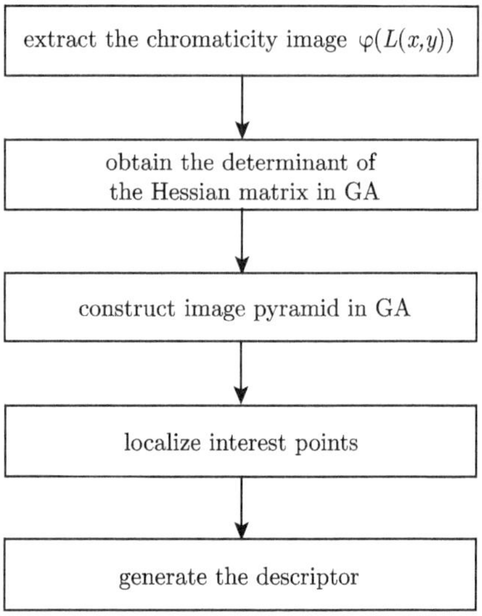

Fig. 4.1 The flow chart of the GA-SURF

4.4 The Proposed GA-SURF Algorithm

In order to verify the performance of the proposed GA-SURF, the experiments are implemented compared with GA-SIFT[21], SURF[22,23] and SURF-DAISY[24] algorithm. Since SURF and SURF-DAISY are designed to deal with gray images or color images which cannot directly extract the features from a multispectral image, only the GA-SURF and GA-SIFT algorithm are compared based on the multispectral image data set. In our experiments, an interest point in the test image is compared to an interest point in the reference image by calculating the Euclidean distance between their descriptor vectors.

4.4.1 Data Set

There are two data sets. One is a multispectral image data set, which can be obtained from http://www.cs.columbia.edu/CAVE/databases/multispectral/. The other is the color image data set, which can be found in http:// www.robots.ox.ac.uk/ ~vgg/data/data-aff.html.

For multispectral images, we randomly choose a multispectral image to evaluate the performance of the GA-SURF. In the experiment, the multispectral image which is selected has 31 bands. Its color image and four single band images are shown in

Fig.4.2. Fig.4.2(a)-(d) show the images of band 1, 2, 3 and 4, respectively. Fig.4.2(e) shows the color image.

(a)Band 1 image (b)Band 2 image (c)Band 3 image (d)Band 4 image (e)Color image

Fig. 4.2 Testing on a multispectral image. (a)-(d) four single band images and (e) the color image

For color images, we choose four transformations: (1) luminance reduction, (2) rotation, (3) image blur, (4) JPEG compression. There are six pictures for every transformation. These pictures have different degrees of the transformation. We select two pictures for each transformation to test the performance of the methods (this experiment used the first and the third image). These pictures used in the experiment are shown in Fig.4.3.

(a)Luminance reduction (b)Rotation

(c)Image blur (d)JPEC compression

Fig. 4.3 Color images for experiments

4.4.2 Evaluation Metrics

Refer to Ref.[21-23], we define the recall-precision as the metric to evaluate the results of different algorithms. The interest points of all the images in the data set are detected and identified. All pairs of interest points from different images are estimated. If the Euclidean distance between the features vectors for a particular pair of interest points from two different images falls below a chosen threshold, this pair is termed a match. A correct-positive is a correct match where the two interest points correspond to the same physical location (using known image transformations for synthetic image deformation tests). A false-positive is an incorrect match where the two interest points come from different physical locations. Among all interest points, if there are two interest points coming from the same physical location, and the two interest points are extracted in different images, we regard this pair of interest points as a positive. The total number of positives for the given data set can be obtained by an affine matrix. The affine matrix is given in the data set. Therefore, we can determine recall and 1-precision as follows:

$$\text{recall} = \frac{\text{number of correct} - \text{positives}}{\text{total number of positives}} \tag{4.31}$$

and

$$\text{precision} = \frac{\text{number of correct} - \text{positives}}{\text{total number of matches(correct or false)}} \tag{4.32}$$

4.4.3 Experimental Results

There are the comparison results of the GA-SURF and the other algorithms on controlled experiments with different transformations.

For multispectral images, Fig.4.4 shows the interest points extracted from a multispectral image with 31 bands by GA-SURF and GA-SIFT. The number of the interest points extracted by GA-SURF which reaches 494, is less than that of GA-SIFT which is 1220. Table 4.1 shows the results of the feature extraction by GA-SURF and GA-SIFT algorithms for multispectral images with different bands, which indicates that the more bands of multispectral images, the more the GA-SURF and GA-SIFT outstanding because the GA-SURF and GA-SIFT can explore both the spectral and geometric information of the multispectral image. The computation times shown in Table 4.1 demonstrate that GA-SURF is faster than GA-SIFT. The reason is that box filters are used instead of the Gaussian second order derivatives. Table 4.2 shows the comparison of the performance of various algorithms for multispectral images ($n = 6$). We can figure it out that the precision and recall of GA-SURF are 0.319 and 0.778, separately, which are both better than that of GA-SIFT, where the precision is 0.189, and the recall is 0.390. As mentioned above, our proposed method can reduce the time for feature computation and matching, and increase simultaneously the robustness.

(a) GA-SURF

(b) GA-SIFT

Fig. 4.4 Interest points extracted from multispectral image ($n = 31$)

Table 4.1 The results of feature extraction from the multispectral images with different bands by various algorithms

number of bands(n)	GA-SURF		GA-SIFT(with parallel pool)	
	interest points	time/s	interest points	time/s
$n = 4$	346	9.57	586	13.72
$n = 5$	361	9.95	641	15.84
$n = 6$	373	10.15	679	17.41

Table 4.2 The performance of various algorithms for multispectral images ($n = 6$)

algorithm	matches (from the images scaled with 50%)	total number of positive (from the images scaled with 50%)	correct-positives (from the images scaled with 50%)	time (from the images scaled with 50%) (includes extraction and matching)
GA-SURF	373	153	119	14.14
GA-SIFT	682	331	129	21.71

For color images, Table 4.3 shows the results of correct-positive interest points extracted from color images by various algorithms, which demonstrate that GA-SURF can extract more correctly matched interest points for images with reduced luminance. In the other distortions, the numbers of the correct matched interest points extracted by the GA-SURF are also comparable with the other algorithms. Fig.4.5 shows the experimental results of features extraction on the color images, where images are distorted by luminance reduction, rotation, image blur and JPEG compression. Fig.4.5(a) demonstrates that the performance of GA-SURF is close to that of GA-SIFT, which is better than that of SURF, SIFT and SURF-DAISY when the image is distorted by reducing the luminance. Fig.4.5(c) shows that GA-SURF approximates or even outperforms other algorithms in the case of the color image blur. In the situations of rotation and JPEG compression, the performance of the GA-SURF is at middle level owing to limitation of bands.

Table 4.3 The results of correct-positive interest points extracted from color images by various algorithms

algorithm	luminance reduction (correct-positive)	rotation (correct-positive)	image blur (correct-positive)	JPEG compression (correct-positive)
GA-SURF	720	273	886	849
GA-SIFT	473	324	871	800
SURF	523	332	1367	1205
SURF-DAISY	382	320	792	1161

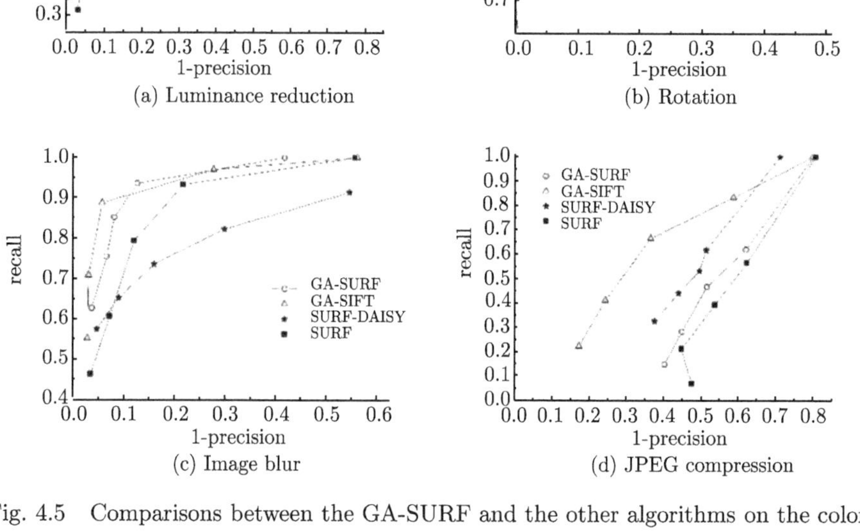

Fig. 4.5 Comparisons between the GA-SURF and the other algorithms on the color images as shown in Fig.4.2

4.5 Conclusions

In this chapter, we propose a new speeded-up robust features based on geometric algebra (GA), denoted as GA-SURF, which uses GA theory to detect and describe the local interest points in multispectral images. The scale space of multispectral images can be obtained by using the scale-varied box filters in GA, and the new Hessian metric is proposed in GA. The proposed scale space and the Hessian metric in GA provide basic conditions for multiscale analysis of multispectral image. Then, the detection of the interest points is completed by using the Hessian matrix in GA. The feature descriptors are generated using Haar wavelet transform in field pixel. Experimental results show that GA-SURF is faster to compute compared with state-of-the-art algorithms while not sacrificing performance. At next steps, we will focus on various applications of multispectral images based on GA-SURF.

References

[1] Shen X, Xu L, Zhang Q, et al. Multi-modal and Multi-spectral Registration for Natural Images. European Conference on Computer Vision. Cham: Springer, 2014: 309-324.

[2] Xie Q, Zhao Q, Meng D, et al. Multispectral images denoising by intrinsic tensor sparsity regularization. Computer Vision and Pattern Recognition. IEEE, 2016: 1692-1700.

[3] Bernabé S, Marpu P R, Plaza A, et al. Spectral-spatial classification of multispectral images using kernel feature space representation. IEEE Geoscience and Remote Sensing Letters, 2014, 11(1): 288-292.

[4] Vivone G, Restaino R, Dalla Mura M, et al. Contrast and error-based fusion schemes for multispectral image pansharpening. IEEE Geoscience and Remote Sensing Letters, 2014, 11(5): 930-934.

[5] Hao Y, Sun Z, Tan T, et al. Multispectral palm image fusion for accurate contact-free palmprint recognition. IEEE International Conference on Image Processing. IEEE, 2008: 281-284.

[6] Saleem S, Sablatnig R. A robust sift descriptor for multispectral images. IEEE Signal Processing Letters, 2014, 21(4): 400-403.

[7] Zhang Q, Cao Z, Hu Z, et al. Joint image registration and fusion for panchromatic and multispectral images. IEEE Geoscience and Remote Sensing Letters, 2015, 12(3): 467-471.

[8] Chaudhuri D, Samal A. An automatic bridge detection technique for multispectral images. IEEE Transactions on Geoscience and Remote Sensing, 2008, 46(9): 2720-2727.

[9] Chen R, Lin X, Ding T. Liveness detection for iris recognition using multispectral images. Pattern Recognition Letters, 2012, 33(12): 1513-1519.

[10] Lowe D G. Object recognition from local scale-invariant features. IEEE International Conference on Computer Vision. IEEE, 1999: 1150-1157.

[11] Lowe D G. Distinctive image features from scale-invariant keypoints. International Journal of Computer Vision, 2004, 60(2): 91-110.

[12] Abdel-Hakim A E, Farag A A. CSIFT: A SIFT descriptor with color invariant characteristics. In Computer Vision and Pattern Recognition, IEEE Computer Society Conference on. IEEE, 2006, 2: 1978-1983.

[13] Zhao Z S, Tian Q J, Wang J Z, et al. Image match using distribution of colorful SIFT. International Conference on Wavelet Analysis and Pattern Recognition. IEEE, 2010: 150-153.

[14] Van De Sande K, Gevers T, Snoek C. Evaluating color descriptors for object and scene recognition. IEEE Transactions on Pattern Analysis and Machine Intelligence, 2010, 32(9): 1582-1596.

[15] Ai D N, Han X H, Ruan X, et al. Color independent components based SIFT descriptors for object/scene classification. IEICE Transactions on Information and Systems, 2010, 93(9): 2577-2586.

[16] Chapman A, Vishne U. Clifford algebras of binary homogeneous forms. Journal of Algebra, 2012, 366: 94-111.

[17] Labunets V. Clifford Algebras as Unified Language for Image Processing and Pattern Recognition. Computational Noncommutative Algebra and Applications. Netherlands: Springer, 2004, 136: 197-225.

[18] Batard T, Saint-Jean C, Berthier M. A metric approach to nD images edge detection with Clifford algebras. Journal of Mathematical Imaging & Vision, 2009, 33(3): 296-312.

[19] Li Y S. A digital image watermarking algorithm based on Clifford algebra. Acta Electronica Sinica, 2008, 36(5): 852-855.

[20] Buchholz S, Sommer G. On Clifford neurons and Clifford multi-layer perceptrons. Neural Networks the Official Journal of the International Neural Network Society, 2008, 21(7): 925-935.

[21] Li Y, Liu W, Li X, et al. GA-SIFT: A new scale invariant feature transform for

multispectral image using geometric algebra. Information Sciences, 2014, 281: 559-572.

[22] Bay H, Ess A, Tuytelaars T, et al. SURF: Speeded Up Robust Features. European Conference on Computer Vision. Berlin Heidelberg: Springer, 2006: 404-417.

[23] Kalia R, Lee K D, Samir B V R, et al. An analysis of the effect of different image preprocessing techniques on the performance of SURF: Speeded Up Robust Features. The Workshop on Frontiers of Computer Vision. IEEE, 2011: 1-6.

[24] Li L, Zic J. Image matching algorithm based on feature-point and DAISY descriptor. Journal of Multimedia, 2014, 9(6): 829-835.

[25] Fan P, Men A, Chen M, et al. Color-SURF: A surf descriptor with local kernel color histograms. IEEE International Conference on Network Infrastructure and Digital Content, 2009. IC-NIDC. IEEE, 2009: 726-730.

[26] Du M, Wang J, Li J, et al. Robot robust object recognition based on fast surf feature matching. Chinese Automation Congress. IEEE, 2013: 581-586.

Chapter 5

Multi-modal Medical Image Registration Based on Feature Spheres in Geometric Algebra

5.1 Introduction

Image registration is to align two images with respect to each other[1]. One of the two is the reference, the other one is the floating image. Medical image registration is essential in medical diagnosis as images of different modalities show very different information. For instance, CT (computed tomography) is best suited for viewing bone injuries, diagnosing lung and chest problems, while MRI (magnetic resonance imaging) is suited for examining soft tissue in ligament and tendon injuries, spinal cord injuries, brain tumors[2]. Anatomic images such as CT and MRI can provide anatomic morphological information of human internal organs, while functional images such as PET (positron emission tomography), SPECT (single photon emission computed tomography) and fMRI (functional MRI) reflect the metabolism of organ functions[3]. Images obtained using different modalities need to be compared to one another and/or combined for analysis and decision making. In general, an image registration algorithm has three components: (1) a deformation model, (2) an objective function and (3) an optimization method[4]. The performance of a registration method largely depends on the deformation model and the objective function. An enormous amount of research has been dedicated to deformable registration due to its potential clinical impact. During the past a few decades, many innovative ideas regrading the three main algorithmic aspects have been proposed. General reviews of these work can be found in Ref.[5]. On the other hand, another line of registration methods is the registration of multi-modal images. These methods generally fall into two categories: intensity-based methods[6–8] and feature-based methods[9–11]. Intensity-based methods mainly use the information of voxel intensities. These methods are robust and accurate, but have large amount of computation. Feature-based methods make use of the similarity of the invariant features extracted from the two images to register. Therefore, feature extraction and analysis has become one of the most important steps in these methods. Among these methods, SURF is widely used due to its efficiency in detecting and describing the features in images[12].

In medical image analysis, the availability of 3D models is of great interest to surgeons because it allows them to have a better understanding of the situation. As human bodies have volumetric structures, feature spheres are more suitable than direct feature points for medical image analysis. Several methods to construct feature spheres from feature points can be found from literatures. For example, Vishwa et al. used a method of Union of Spheres (UoS) to represent the volumetric data and built 3D models of objects[13]. To solve the problem of heavy computation in UoS, marching cubes algorithm was proposed to reduce the number of spheres[14]. However, this method needed a preprocessing stage to segment the objects of interest based on texture information.

In this work, we propose a novel method, GA-SURF, which incorporates the geometric algebra (GA) theory into SURF algorithm, to detect features from multimodal images. With the advantages of the GA theory, GA-SURF algorithm can process high dimensional information. To speedup the process of image registration, we propose a fast method to construct feature spheres from feature points without the need of image segmentation. Briefly, we start with two feature points and draw a circle with them on the diameter. Then we greedily search for the next point in the remaining candidates to construct a sphere with the circle on the equator. The step repeats until all the feature points are processed. After we have all the feature spheres, we find the correspondence between them to register the images.

The rest of the chapter is organized as follows. Section 5.2 describes our novel approach, including its theoretical basis and design consideration. Experimental results are presented in Section 5.3. Section 5.4 concludes the chapter.

5.2 Method

5.2.1 SURF Algorithm

In this section, we briefly introduce the SURF algorithm, which has been widely used in computer vision applications[12]. Like the algorithm of scale invariant feature transform (SIFT)[15], SURF has two major steps, i.e. feature detection and descriptor.

1. Feature Point Detection

Hessian matrix is the core of SURF algorithm. It constructs the Hessian matrix from the image after smoothing with Gaussian filters and uses the integral image to speedup the box type convolutions. Suppose $I_{\Sigma}(\boldsymbol{x})$ represents the sum of all pixel indensities in the input image I within a rectangular region formed by the origin and a location $\boldsymbol{x} = (x, y)$.

$$I_{\Sigma}(\boldsymbol{x}) = \sum_{i=0}^{i \leqslant x} \sum_{j=0}^{j \leqslant y} I(i, j) \tag{5.1}$$

With the integral image, calculating the sum of the intensities in any rectangular area can be easily done by simple additions and subtractions.

Constructing the scale space is a very important step in SURF, which allows the algorithm to observe the image from different aspect. The scale space is implemented as an image pyramid, which, like SIFT, is divided into octaves by using box filters of different sizes applied on the original image. Each octave is further divided into a constant number of scale levels[16]. SURF detects the feature points based on local maximum determinant of the Hessian matrix. For a given point $\boldsymbol{x} = (x, y)$ in an image I, the Hessian matrix $H(\boldsymbol{x}, \sigma)$ at $\boldsymbol{x}$ with scale σ is defined as

$$H(\boldsymbol{x}, \sigma) = \begin{pmatrix} L_{xx}(\boldsymbol{x}, \sigma) & L_{xy}(\boldsymbol{x}, \sigma) \\ L_{xy}(\boldsymbol{x}, \sigma) & L_{yy}(\boldsymbol{x}, \sigma) \end{pmatrix} \tag{5.2}$$

Where $L_{xx}(\boldsymbol{x}, \sigma)$ is the convolution of the Gaussian second order derivative $\dfrac{\partial^2 g(\sigma)}{\partial x^2}$ with image I at point $\boldsymbol{x}$. $L_{xy}(\boldsymbol{x}, \sigma)$ and $L_{yy}(\boldsymbol{x}, \sigma)$ are similar. SURF adopts the convolution of the box filter with the integral image to approximate L_{xx}, L_{xy} and L_{yy}, denoted by D_{xx}, D_{xy} and D_{yy} respectively.

The determinant of the approximated Hessian is

$$\det(H_{\text{approx}}) = D_{xx}D_{yy} - (wD_{xy})^2 \tag{5.3}$$

where w is a weight, usually set to 0.9 in practice. When a point has greater response than its surrounding neighbors, it is a feature point.

2. Feature Point Descriptor

Feature point orientation is obtained by counting the Haar wavelet responses in x and y direction in a circular region with radius of $6s$ centered at the point of interest. The Haar wavelet size is $4s$, where s is the scale at which the interest point is detected. The dominant orientation is estimated by calculating the sum of all responses within a sliding orientation window covering an angle of $\frac{\pi}{3}$.

To compute the feature vector, the square area with side length of $20s$ centered around the interest point is divided into 4×4 sub-regions. In each sub-region, dx and dy are obtained by calculating the Haar wavelet response in horizontal and vertical direction respectively. Each sub-region can be denoted $v = \left(\sum dx, \sum dy, \sum |dx|, \sum |dy|\right)$. Therefore we have a 64-dimensional vector in the 4×4 sub-regions.

5.2.2 The Basics of Geometric Algebra

Geometric algebra (GA), also named Clifford algebra, has important applications in theoretical physics and computer vision as it can improve the computational efficiency[17]. We exploit a set of orthogonal basis $(e_1, e_2, \cdots, e_n)$ to define geometric algebra in n-dimensional $\mathbb{G}_n$. Some of the important properties about the basis are listed as follows.

$$e_ie_j = -e_je_i, \quad i \neq j, \quad i, j \in [1, n] \tag{5.4}$$

$$e_i^2 = 1, \quad i \in [1, n] \tag{5.5}$$

$$\begin{aligned} e_i \cdot e_j &= \frac{1}{2}(e_ie_j + e_je_i) \\ &= \delta_{ij}, \quad i, j \in [1, n] \end{aligned} \tag{5.6}$$

If $a \in \mathbb{G}_n$, a can be written as

$$a = a_0 + \sum_{i=0}^{n} a_i e_i \tag{5.7}$$

where $a_0, a_1, \cdots, a_n \in \mathbb{R}$. In GA, the geometric product is expressed in terms of inner product and outer product.

$$ab = a \cdot b + a \wedge b = \frac{1}{2}(ab + ba) + \frac{1}{2}(ab - ba) \tag{5.8}$$

where the inner product $a \cdot b = \frac{1}{2}(ab + ba)$ and the outer product $a \wedge b = \frac{1}{2}(ab - ba)$. The outer product is anti-commutative, namely, $a \wedge b = -b \wedge a$. So the geometric product is neither fully symmetric, nor fully anti-symmetric.

$$ba = b \wedge a + b \cdot a = -a \wedge b + a \cdot b \tag{5.9}$$

We know a complex number has the real part and imaginary part. Similarly, ab is composed of a scalar and a bivector. A k-blade is generally defined as an outer product of k linearly independent vectors.

For example, the outer product of two vectors is referred to as a 2-blade.

The linear combination of blades makes up a multivector which can be written as

$$M = a_0 + \sum_{i_1} a_{i_1} e_{i_1} + \sum_{i_1 \neq i_2} a_{i_1 i_2} e_{i_1 i_2} + \cdots + \sum_{i_1 \neq i_2 \neq \cdots \neq i_n} a_{i_1 i_2 \cdots i_n} e_{i_1 i_2 \cdots i_n} \tag{5.10}$$

where $a_i \in \mathbb{R}$.

5.2.3 The GA-SURF Algorithm

As mentioned previously, GA-SURF incorporates the geometric algebra (GA) theory into SURF framework to detect features. In this section, we explain the details in two parts, i.e. feature point detection and feature descriptor in GA space.

1. Feature Point Detection in GA Space

The integral image, box filter convolution used in conventional SURF algorithm are not applicable in GA space. Therefore, we need redefine the integral images, box filters and the convolution in GA space.

The integral image in GA space

According to Eq. (5.1), we have the sum of the intensities $I_\Sigma(a,b)$ in a hyper-box that ranges from $\boldsymbol{a}=(a_1,a_2,\cdots,a_N)$ to $\boldsymbol{b}=(b_1,b_2,\cdots,b_N)$ as

$$I_\Sigma(\boldsymbol{a},\boldsymbol{b})=\sum_{i_1=a_1}^{b_1}\sum_{i_2=a_2}^{b_2}\cdots\sum_{i_N=a_N}^{b_N}I(i_1,i_2,\cdots,i_N) \tag{5.11}$$

where $I(i_1,i_2,\cdots,i_N)$ is an N-dimensional image. From Eq. (5.1) and Eq. (5.11), we have

$$I_\Sigma(\boldsymbol{b})=I_\Sigma(\boldsymbol{0},\boldsymbol{b})=\sum_{i_1=0}^{b_1}\sum_{i_2=0}^{b_2}\cdots\sum_{i_N=0}^{b_N}I(i_1,i_2,\cdots,i_N) \tag{5.12}$$

The integral image in GA space is N-dimensional.

The convolution in GA space

For N-dimensional vectors, Bulow[18] defined the convolution in GA space as

$$\begin{aligned}v_N(r_1,r_2,\cdots,r_N)=&\int\int\cdots\int Q_N(\xi_1,\xi_2,\cdots,\xi_N)\\ &V(r_1-\xi_1,r_2-\xi_2,\cdots,r_N-\xi_N)\mathrm{d}\xi_1\mathrm{d}\xi_2\cdots\mathrm{d}\xi_N\end{aligned} \tag{5.13}$$

where $Q_N(\xi_1,\xi_2,\cdots,\xi_N)$ and $V(r_1-\xi_1,r_2-\xi_2,\cdots,r_N-\xi_N)$ are two N-dimensional multivectors, and the product of $Q_N(\xi_1,\xi_2,\cdots,\xi_N)$ and $V(r_1-\xi_1,r_2-\xi_2,\cdots,r_N-\xi_N)$ is the geometric product. The convolution of two discrete multivectors in GA space is defined as

$$\begin{aligned}&v_N(r_1,r_2,\cdots,r_N)\\ =&\sum_{i_1=0}^{\infty}\sum_{i_2=0}^{\infty}\cdots\sum_{i_N=0}^{\infty}Q_N(i_1,i_2,\cdots,i_N)V(r_1-i_1,r_2-i_2,\cdots,r_N-i_N)\end{aligned} \tag{5.14}$$

However, as the geometric product is neither fully symmetric, nor fully anti-symmetric (see Eq.(5.8) and Ref.[9]), we have the right-side and left-side convolutions with different order of multiplication of $Q(\ )$ and $V(\ )$ in Eq. (5.14). The right-side convolution in GA space is defined as

$$F=I\otimes h=\sum_{j_1=0}^{\infty}\sum_{j_2=0}^{\infty}\cdots\sum_{j_N=0}^{\infty}I(i_1-j_1,i_2-j_2,\cdots,i_N-j_N)h(j_1,j_2,\cdots,j_N) \tag{5.15}$$

where $I(i_1,i_2,\cdots,i_N)$ is an N-dimensional image, $h(j_1,j_2,\cdots,j_N)$ is a function in GA space, and the product of $I(i_1-j_1,i_2-j_2,\cdots,i_N-j_N)h(j_1,j_2,\cdots,j_N)$ in

Eq.(5.15) is the geometric product. The left-side version is defined as

$$F = h \otimes I = \sum_{j_1=0}^{\infty} \sum_{j_2=0}^{\infty} \cdots \sum_{j_N=0}^{\infty} h(j_1, j_2, \cdots, j_N) I(i_1 - j_1, i_2 - j_2, \cdots, i_N - j_N) \tag{5.16}$$

The box filter in GA space

Box filters are used in SURF to speedup the convolution. According to Eq.(5.2), we have the following approximate Hessian matrix in N-dimensional GA space from the convolution of the integral image in GA space with the N-dimensional (hyper)-cuboid-filter.

$$H_{\text{approx}}(\boldsymbol{x}, \sigma) = \begin{bmatrix} D_{i_1 i_1}(\boldsymbol{x}, \sigma) & \ldots & D_{i_1 i_N}(\boldsymbol{x}, \sigma) \\ D_{i_2 i_1}(\boldsymbol{x}, \sigma) & \ldots & D_{i_2 i_N}(\boldsymbol{x}, \sigma) \\ \vdots & & \vdots \\ D_{i_N i_1}(\boldsymbol{x}, \sigma) & \ldots & D_{i_N i_N}(\boldsymbol{x}, \sigma) \end{bmatrix} \tag{5.17}$$

where each matrix element is the convolution of the hyper-cuboid filters with the image I, and $i_1, \cdots, i_N$ are the dimension indexes. In GA space, an interest point in an image has the maximum determinant of the approximate Hessian matrix among its neighboring pixels.

2. The Descriptor in GA Space

As mentioned earlier, traditional SURF algorithm uses Haar wavelets to construct the feature descriptors. In our GA-SURF algorithm, Haar filters are represented by combinations of rectangle filters in N-dimensional GA space. The samples are split into b bins per dimension, resulting in b^N bins for N dimensions. For each bin, we can obtain the feature vector as

$$v = \left(\sum d_1, \cdots, \sum d_N, \sum |d_1|, \cdots, \sum |d_N|\right) \tag{5.18}$$

where d_i is the Haar wavelet response at the i-th dimension. The final descriptor is generated by concatenating the feature vectors from all bins. Thus, the final descriptor has a dimension of $l = 2Nb^N$.

5.2.4 Applying GA-SURF to the Medical Images

In this section, we apply the GA-SURF algorithm to detect features from medical images. Before that, we represent the CT and MRI images in GA space. Usually, medical tomography images are consisted of many slices. A CT or MRI image $I(x, y, z)$ can be represented as

$$I(x, y, z) = \sum_{l=1}^{L} S_l(x, y, z_l) \tag{5.19}$$

where $I(x, y, z)$ is the l-th slice of $I(x, y, z)$ along z dimension, L is the total slices of the image. A more general form of Eq.(5.19) is

$$I(x) = \sum_{i=1}^{L} \text{proj}(S_l, A)$$

where $\text{proj}(S_l, A)$ is a projection of S_l in any plane A, and S_l is the l-th slice.

In Eq.(5.19), the slices are projected to the surface of X-Y plane.

1. Feature Point Detection

With Eq.(5.19), we have the following steps to detect feature points

Step 1 Calculate the integral image

From Eq.(5.12), the integral image $IP(x)$ of a medical tomography image I at a point $\boldsymbol{x} = (a, b, c)$ can be written as

$$I_{\Sigma}(a, b, c) = \sum_{i_1=0}^{a} \sum_{i_2=0}^{b} \sum_{i_3=0}^{c} I(i_1, i_2, i_3) \tag{5.20}$$

Step 2 Construct the scale space

In 3D discrete space, Eq.(5.14) becomes

$$v_N(r_1, r_2, r_3) = \sum_{i_1=0}^{\infty} \sum_{i_2=0}^{\infty} \sum_{i_3=0}^{\infty} Q_N(i_1, i_2, i_3) V(r_1 - i_1, r_2 - i_2, r_3 - i_3) \tag{5.21}$$

The right-side convolution of Eq.(5.15) is

$$F = I \otimes h = \sum_{j_1=0}^{\infty} \sum_{j_2=0}^{\infty} \sum_{j_3=0}^{\infty} I(i_1 - j_1, i_2 - j_2, i_3 - j_3) h(j_1, j_2, j_3) \tag{5.22}$$

where $I(i_1, j_2, j_3)$ is the integral image defined in Eq.(5.20) and $h(j_1, j_2, j_3)$ is the cuboid filter.

Similarly, from Eq.(5.16), the left-side convolution is

$$F = h \otimes I = \sum_{j_1=0}^{\infty} \sum_{j_2=0}^{\infty} \sum_{j_3=0}^{\infty} h(j_1, j_2, j_3) I(i_1 - j_1, i_2 - j_2, i_3 - j_3) \tag{5.23}$$

Similarly to SURF algorithm, the scale space is constructed by the convolution between the integral image and the cuboid filters of different sizes, and is divided into octaves. Each octave is further divided into a constant number of scale levels.

Step 3 Locating the feature points

GA-SURF detects the feature point based on the local maximum determinant of the Hessian matrix. The 3D version of the approximated Hessian matrix defined in Eq.(5.17) can be written as

$$H_{\text{approx}}(\boldsymbol{x},\sigma)=\begin{bmatrix} D_{i_1i_1}(\boldsymbol{x},\sigma) & D_{i_1i_2}(\boldsymbol{x},\sigma) & D_{i_1i_3}(\boldsymbol{x},\sigma) \\ D_{i_2i_1}(\boldsymbol{x},\sigma) & D_{i_2i_2}(\boldsymbol{x},\sigma) & D_{i_2i_3}(\boldsymbol{x},\sigma) \\ D_{i_3i_1}(\boldsymbol{x},\sigma) & D_{i_3i_2}(\boldsymbol{x},\sigma) & D_{i_3i_3}(\boldsymbol{x},\sigma) \end{bmatrix}$$

When a point has greater determinant of the approximate Hessian matrix than its surrounding neighbors, it is treated as a feature point.

2. Feature Point Descriptor

After the detection of feature points, we can calculate the feature descriptors in the following similar steps to SURF.

Step 1 Calculate the orientation of feature point

Orientation is obtained by counting the Haar wavelet responses in x and y direction in a circular region centered at the interest point with radius of $6s$, where s is the scale and the Haar wavelet size is $4s$.

Step 2 Calculate the feature descriptor

In 3D, the image is sampled on a regular grid around an interest point. The samples are split into b bins per dimension, resulting in b^3 bins. For each bin, a feature vector can be obtained as

$$v=\left(\sum d_1,\sum d_2,\sum d_3,\sum |d_1|,\sum |d_2|,\sum |d_3|\right)$$

where d_1,d_2,d_3 are the Haar filter response in dimension of x,y and z respectively.

The final descriptor is generated by concatenating the feature vectors from all bins.

5.2.5 Construct Feature Spheres

In previous sections, we have explained the GA-SURF on how to detect the feature points and calculate the feature descriptors. As human body have volumetric structures, feature spheres are more suitable for surgeons in medical image analysis. Moreover, because feature spheres are composed of a set of feature points, matching the feature spheres is more accurate and efficient than directly matching the feature points.

In our new algorithm to construct the feature spheres, there are two predefined thresholds d_1 and d_2. Empirically, $d_1 = 1$mm and $d_2 = 3$mm. Fig.5.1 and Fig.5.2 are used to help explain the following steps.

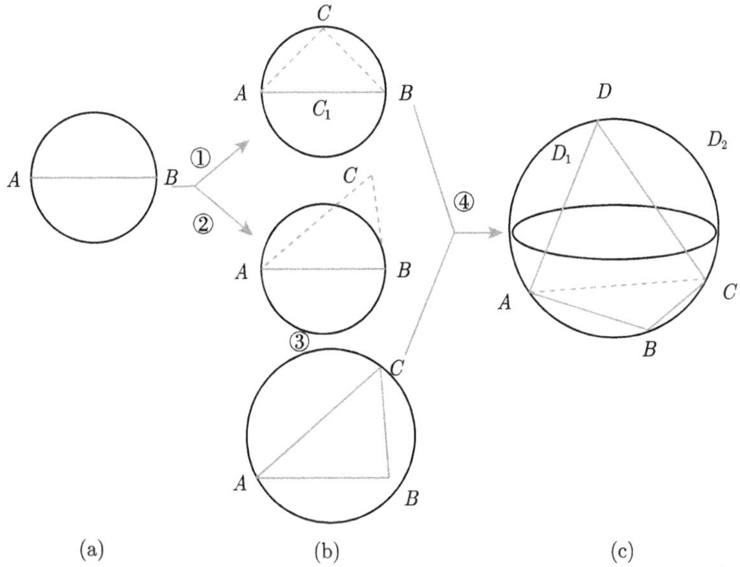

Fig.5.1 Construct the feature spheres

1) Randomly pick up one point A, draw a sphere centered at A with radius d_2. If there is no point within the sphere, draw a sphere with radius d_1. The sphere is done and go to next point.

2) If there is one point B found inside the sphere, draw a circle with AB as the diameter, and decide if there are any other points covered by the circle, i.e points are on or inside the circle.

3) If additional points are covered by the circle, the circle is valid and we go to case ④ in Fig. 5.1. Case ① shows C is on the circle and C_1 is inside the circle. If there is no point covered by the circle (case ②), we need enlarge the circle until the diameter reaches d_2 (case ③). We abandon A and B if no such points found after

the reaches the maximum.

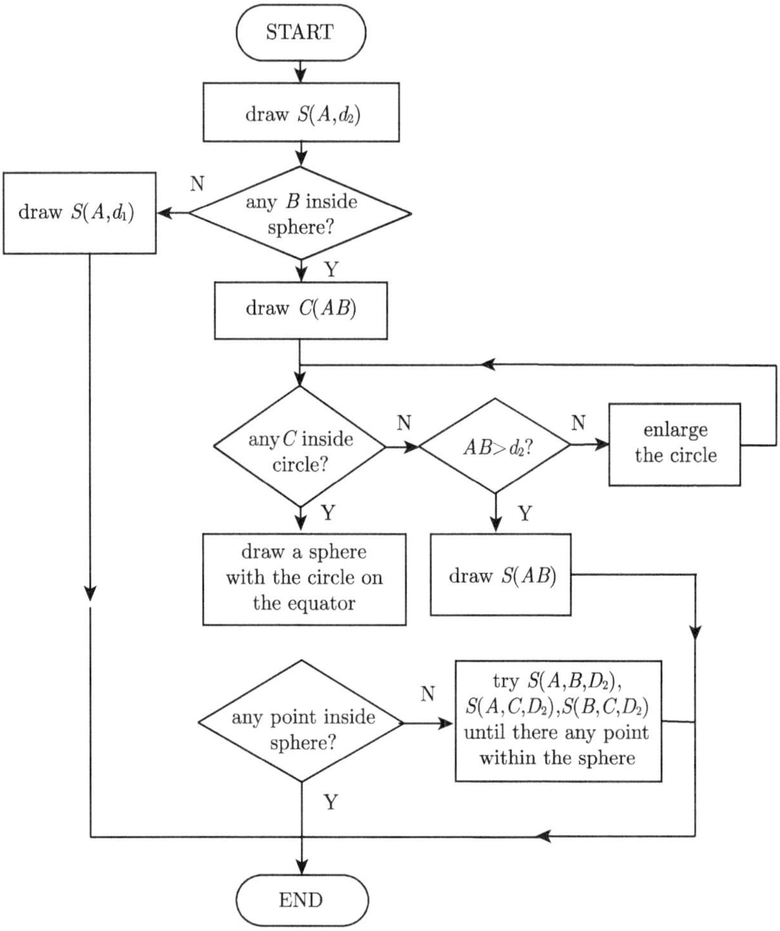

Fig.5.2 Flowchart of constructing the feature spheres. $S(A, d_1)$ means a sphere center at point A with radius of d_1, similar for $S(A, d_2)$. $C(\overline{AB})$ mean a circle with $\overline{AB}$ as the diameter. $S(A, B, D_2)$ means a sphere with a circumcircle of the triangle ABD_2 on the equator, similar for $S(A, C, D_2)$ and $S(B, C, D_2)$

4) With a circle drawn from points A, B and C, we go to case ④ to draw a sphere with the circle on the equator. Decide if any points other than the plane ABC covered by the sphere. For example, in the figure, D is on the sphere and D_1 is inside and D_2 is outside.

5) If there are some points covered by the sphere, then the sphere is valid and done. Otherwise, take D_2 for instance, we need to try the circumcircle of the triangle

ABD_2, BCD_2 and ACD_2 until we find a valid circle which has the diameter less than d_2. If a valid circle is found, we repeat ④ to draw a sphere.

5.2.6 Conformal Geometric Algebra

Conformal geometric algebra (CGA) as an unified homogeneous algebraic framework provides an effective mathematical tool for geometric representation and computation[19]. In general, traditional Euclidean geometric algebra has three basis vectors: e_1, e_2, e_3. CGA adds two basis vectors e_0 and e_∞, which represent the origin and the infinite point respectively. In addition, they are null vectors

$$e_0^2 = e_\infty^2 = 0 \tag{5.24}$$

The inner product of e_0 and e_∞ is

$$e_0 \cdot e_\infty = -1 \tag{5.25}$$

In addition, unit pseudoscalar is defined as

$$E = e_0 \wedge e_\infty \tag{5.26}$$

In CGA, spheres are the most basic geometric entities because other geometric entities can be derived from the spheres. A sphere is expressed as

$$S = c + \frac{1}{2}(c^2 - r^2)e_\infty + e_0 \tag{5.27}$$

where c is the center of the sphere and r is the radius.

1. Motion in Conformal Geometric Algebra

In general, the transformation of rigid body has three types: rotation, translation and dilation. In CGA, rotors, translators and dilators are defined as follows

$$R = \mathrm{e}^{(\frac{\theta}{2})a} \tag{5.28}$$

$$T = \mathrm{e}^{e_\infty t/2} \tag{5.29}$$

$$D = \mathrm{e}^{(\frac{-\log\lambda}{2})\wedge E} \tag{5.30}$$

where the bivector a is the dual to the rotation axis, θ is the angle of rotation, t is the amount of translation, and λ is the scaling factor. By multiplying from the left with the rotor R, and from the right with its inverse R, the rotation of an entity X can be expressed as $X' = RX\bar{R}$. Similarly, the translation of X is $X' = XT\bar{T}$. The dilator works differently as it only works at the origin. Therefore, an entity must be translated to the origin before being applied the dilator.

$$X' = DT_0X\overline{T_0}\bar{D} \tag{5.31}$$

where $\bar{D}$ is the inverse of D and T_0 is the corresponding translator to the origin. Fig.5.3 shows an example of the correspondence of two feature spheres S_1 and S_2. From Eq.(5.27), the sphere center c and radius r can be represented as

$$c = (S \wedge E) \cdot E \tag{5.32}$$

$$r = \sqrt{(-S \cdot S)} \tag{5.33}$$

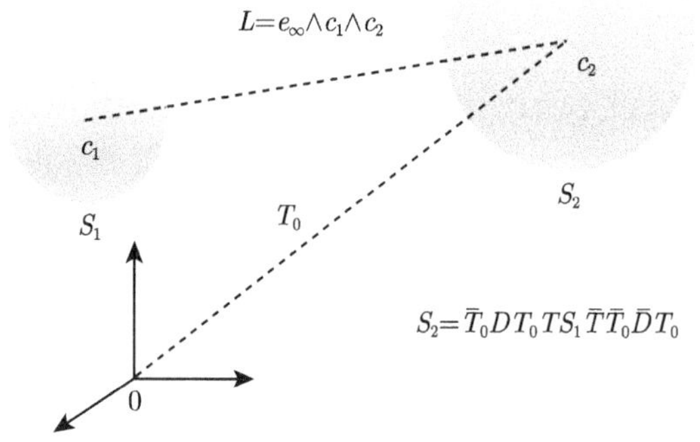

Fig.5.3 Transformation from S_1 to S_2

Once we have the centers c_1, c_2 and radius r_1, r_2 of the two feature spheres, we can obtain the line connecting the two centers by $L = e_\infty \wedge c_1 \wedge c_2$ and its direction n_L. Then, we can compute the translator T and T_0 based on Eq.(5.23), $t = n_L$, and $t_0 = c_2$. From Eq.(5.32) and $\lambda = \dfrac{r_2}{r_1}$, we can get the dilator. Finally, the transformation from S_1 to S_2 is $\bar{T}_0DT_0TS_1\bar{T}\bar{T}_0\bar{D}T_0$. Once the correspondence of feature spheres is obtained, images can be registered.

2. Match Feature Spheres

The descriptor of a feature sphere is the weighted sum of the feature descriptors of those points inside (or on) the sphere. We calculate the distance between feature spheres based on the distance between their descriptors using inner product defined in CGA. With that, we can find the correspondence of the spheres in the reference and floating image.

5.3 Results

To test our method, we carry out three different experiments. All the data are downloaded from "retrospective image registration evaluation" (RIRE)[20]. The results are also evaluated by RIRE.

Experiment 1

In this experiment, we register a CT image and a mr-PD image from data patient_001 in RIRE project. Fig.5.4(a) and (b) shows the CT and mr-PD image respectively. The size of the CT image data is 512×512×28. Here the data size should be read as 28 images of size 512×512. The same for the data sizes in the rest of the chapter. The size of the mr-PD image data is 256 × 256 × 26. In Fig.5.4(a), the top-left corner is the three-dimensional diagram, and the top-right corner, bottom-left, bottom-right are the cross section, sagittal section and coronal respectively. Fig.5.4(b) is similar. Fig.5.5 shows one correspondence between the two spheres from CT and mr-PD images, which is calculated using the method described above. To compare and highlight the registration effect, we show the fusion before and after registration using pseudo-colors in Fig.5.6(a) and (b) respectively. From the figure,

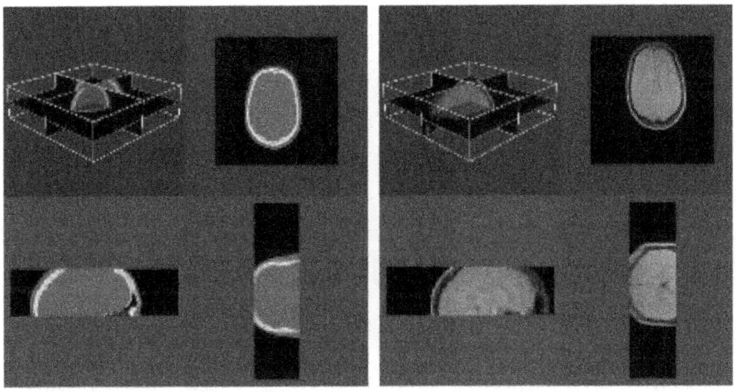

(a) (b)

Fig.5.4 CT and mr-PD image from data patient_001. (a) CT image, (b) mr-PD image. For subfigure (a) and (b), the top-left corner is the 3D diagram, top-right is the cross section, bottom-left is the sagittal section and bottom-right is the coronal

we can see that the registration of CT and mr-PD is excellent. We also show the evaluation provided by RIRE in Table 5.1. The mean, max and median error are 2.345mm, 3.502mm and 2.176mm, respectively. The range of these errors is 2 ~ 4mm, which means our method has achieved high registration accuracy.

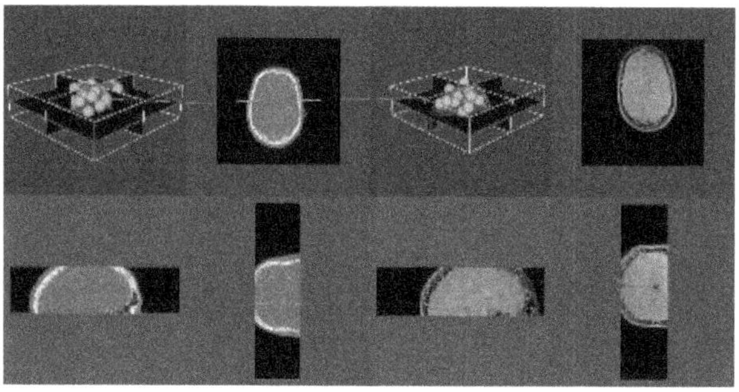

Fig.5.5 One correspondence of two feature spheres

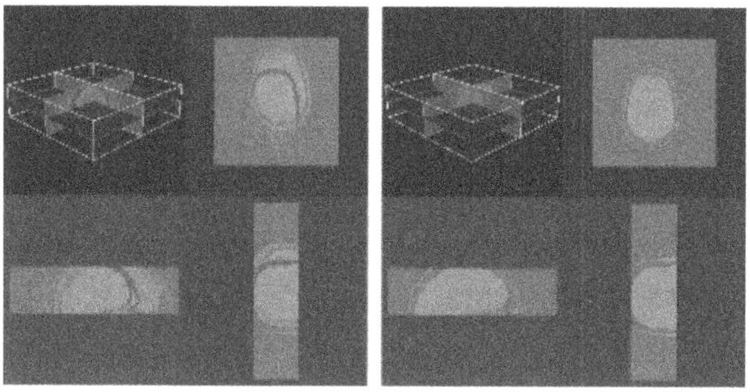

(a) Fusion before registration (b) Fusion after registration

Fig.5.6 Image fusion

Table 5.1 RIRE evaluation of our results

volume	VOI1	VOI2	VOI3	VOI4	VOI5
error/nm	2.157	2.075	2.90	3.502	2.39
volume	VOI6	VOI7	VOI8	VOI9	VOI10
error/nm	2.868	2.195	1.672	1.942	1.751
mean error			2.345mm		
median error			2.176mm		
max error			3.502mm		

Experiment 2

Experiment 2 is the registration between a CT and a mr-T1 image of patient_001. Fig.5.7 (a) and (b) shows the two images respectively. The respective image size is

512×512×28 and 256 × 256 × 26. Sections in Fig.5.7 are similar to Fig.5.4. Fig.5.8 shows one correspondence between the two feature spheres in CT and mr-T1 image. Fig.5.9(a) and (b) shows the fusion before and after registration. And Table 5.2 shows the RIRE evaluation of the results. The mean, max and median error are 2.260mm, 3.316mm and 2.158mm, respectively. The range of these errors is also 2 ~ 4mm.

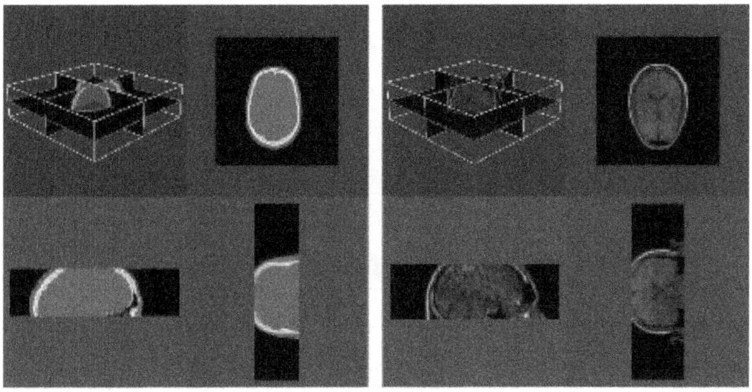

(a) (b)

Fig.5.7 CT and mr-T1 image of data patient_001. (a) CT image, (b) mr-T1 image. For subfigure (a) and (b), the top-left corner is the 3D diagram, top-right is the cross section, bottom-left is the sagittal section and bottom-right is the coronal

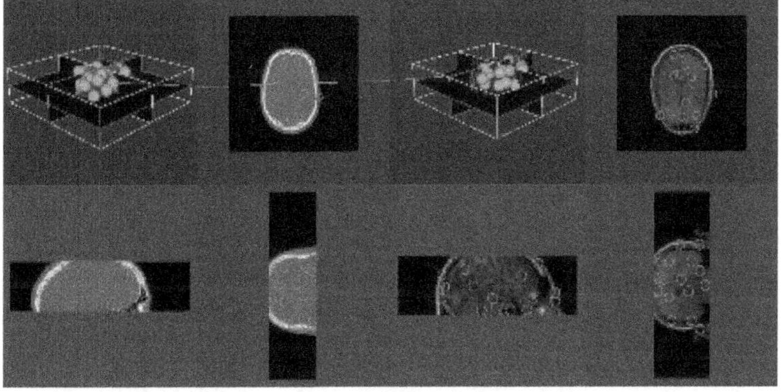

Fig.5.8 One correspondence of two feature spheres

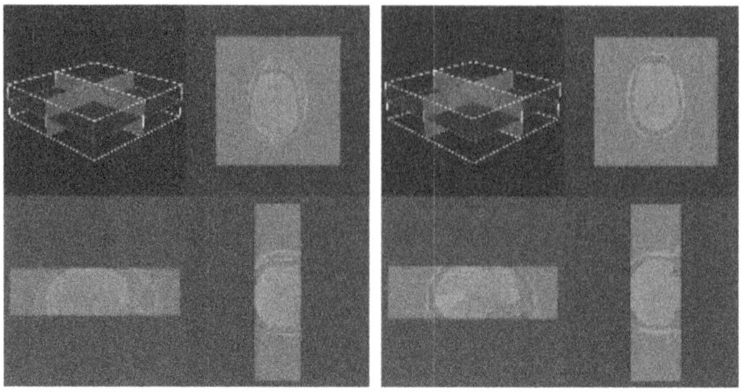

(a) Before registration (b) After registration

Fig.5.9 Image fusion

Table 5.2 RIRE evaluation of our results

volume	VOI1	VOI2	VOI3	VOI4	VOI5
error/nm	1.869	1.847	2.762	3.316	2.163
volume	VOI6	VOI7	VOI8	VOI9	VOI10
error/nm	2.702	2.153	2.21	1.783	1.792
mean error			2.260mm		
median error			2.158mm		
max error			3.316mm		

Experiment 3

To prove the robustness of our algorithm, experiment 3 is carried out to register a CT and a mr-PD_rectified image of patient_005. Fig.5.10 (a) and (b) shows the images of CT and mr-PD_rectified respectively. Their image sizes are $512 \times 512 \times 33$ and $256 \times 256 \times 26$. Similarly, Fig.5.11 shows one correspondence of two spheres in CT and mr-PD_rectified. Fig.5.12 (a) and (b) shows the fusion before and after registration. Again, the registration shown in the figure is very good. Table 5.3 shows the RIRE evaluation of this experimental results. The mean, max and median error are 2.881mm, 3.436mm and 2.888mm, respectively. The range of these errors is $2 \sim 4$mm, which has reached sub-pixel registration accuracy. Fig.5.13 shows an example in 3D effect. From the figure, we can see that the two figures match perfectly after registration.

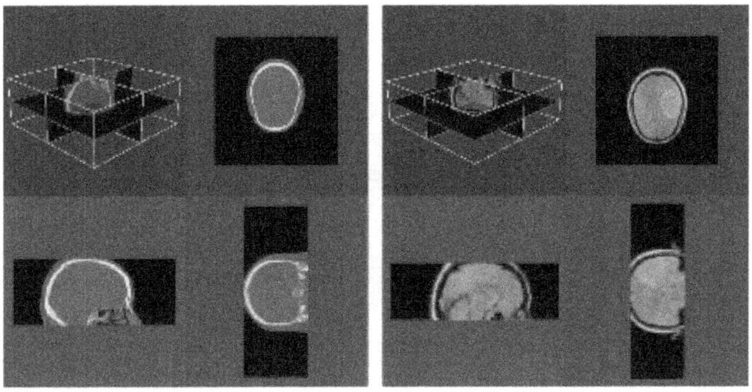

(a) (b)

Fig.5.10 CT and mr-PD_rectified images of data patient_005. (a) CT, (b) mr-PD_rectified. For subfigure (a) and (b), the top-left corner is the 3D diagram, top-right is the cross section, bottom-left is the sagittal section and bottom-right is the coronal

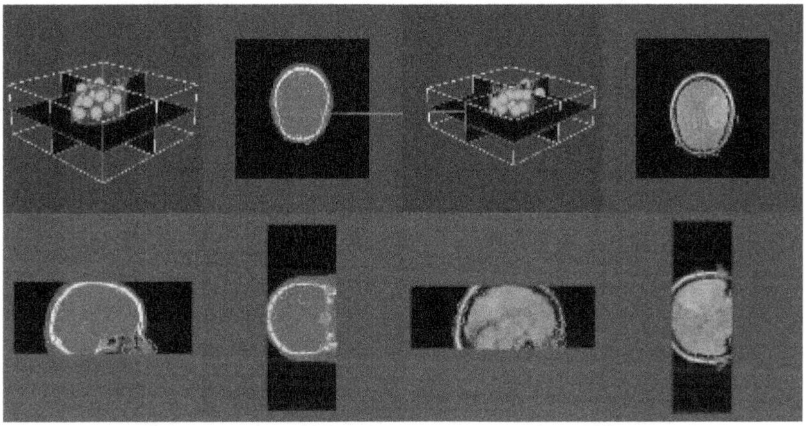

Fig.5.11 One correspondence of two feature spheres

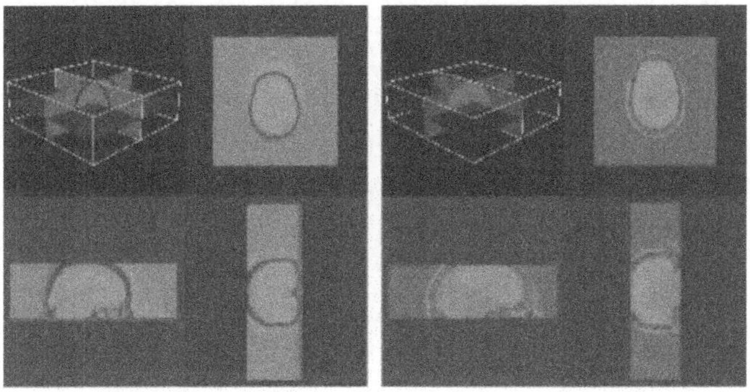

(a) Before registration (b) After registration

Fig.5.12 Fusion

Table 5.3 RIRE evaluation of our results

volume	VOI1	VOI2	VOI3	VOI4	VOI5
error/nm	2.888	2.888	2.215	3.047	2.728
volume	VOI6	VOI7	VOI8	VOI9	VOI10
error/nm	3.352	2.15	3.27	2.838	3.436
mean error			2.881mm		
median error			2.888mm		
max error			3.436mm		

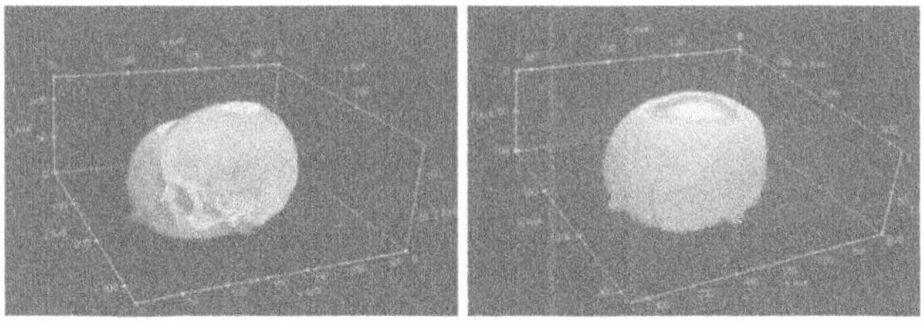

(a) Before registration (b) After registration

Fig.5.13 3D images for experiment 3

5.4 Conclusions

In this chapter, we present a new algorithm to model volumetric data and register multi-modal medical images. We detect the feature points using the proposed new algorithm, GA-SURF, which incorporates the geometric algebra theory into the SURF framework. With the features detected, we apply our fast greedy algorithm to construct the feature spheres formulated in conformal geometric algebra (CGA). The last step is to find the correspondence of feature spheres in two images and register the images based on the correspondences. We test our method using data of different modality from different patients. The experimental results evaluated by RIRE have shown that our method can achieve high registration accuracy with max error less than 4mm.

References

[1] El-Gamal F E-Z A, Elmogy M, Atwan A. Current trends in medical image registration and fusion. Egyptian Informatics Journal, 2016, 17(1): 99-124.

[2] Rui W, Minglu L. An overview of medical image registration. in Computational Intelligence and Multimedia Applications, 2003. ICCIMA, 2003. Proceedings Fifth International Conference on. IEEE, 2003: 385-390.

[3] Oliveira F P, Tavares J M R. Medical image registration: A review. Computer Methods in Biomechanics and Biomedical Engineering, 2014, 17(2): 73-93.

[4] Sotiras A, Davatzikos C, Paragios N. Deformable medical image registration: A survey. IEEE Transactions on Medical Imaging, 2013, 32(7): 1153-1190.

[5] Brown L G. A survey of image registration techniques. ACM Computing Surveys (CSUR), 1992, 24(4): 325-376.

[6] Alam M M, Howlader T, Rahman S M. Entropy-based image registration method using the curvelet transform. Signal, Image and Video Processing, 2014, 8(3): 491-505.

[7] Kim J, Fessler J A. Intensity-based image registration using robust correlation coefficients. IEEE Transactions on Medical Imaging, 2004, 23(11): 1430-1444.

[8] Xu L, Wan J W. Real-time intensity-based rigid 2D-3D medical image registration using rapidmind multi-core development platform. in Engineering in Medicine and Biology Society, 2008. EMBS 2008. 30th Annual International Conference of the IEEE. IEEE, 2008: 5382-5385.

[9] Smeets D, Keustermans J, Hermans J, et al. Feature-based piecewise rigid registration in 2D medical images. in Biomedical Imaging (ISBI), 2012 9th IEEE International Symposium on. IEEE, 2012: 696-699.

[10] Hsu L Y, Loew M H. Fully automatic 3D feature-based registration of multi-modality medical images. Image and Vision Computing, 2001, 19(1): 75-85.

[11] Hahn D A, Wolz G, Sun Y, et al. A practical salient region feature based 3D multimodality registration method for medical images. Proceedings of the Medical Imaging: Image Processing, 2006, 6144: 61442P.

[12] Bay H, Tuytelaars T, VanGool L. SURF: Speeded Up Robust Features. Computer Vision–ECCV 2006: 404-417.

[13] Ranjan V, Fournier A. Union of Spheres (UoS) model for volumetric data. in Proceedings of the Eleventh Annual Symposium on Computational Geometry. ACM, 1995: 402-403.

[14] Jorge R R, Eduardo B C. Medical image segmentation, volume representation and registration using spheres in the geometric algebra framework. Pattern recognition, 2007, 40(1): 171-188.

[15] Lowe D G. Object recognition from local scale-invariant features. in Computer Vision, 1999. The Proceedings of the Seventh IEEE International Conference on, IEEE, 1999,

2: 1150-1157.

[16] Christopher E. Notes on the opensurf library. University of Bristol Technical Paper, 2009.

[17] Perwass C, Edelsbrunner H, Kobbelt L, et al. Geometric Algebra with Applications in Engineering. Berlin, Germany: Springer, 2009: 20.

[18] Bülow T. Hypercomplex spectral signal representations for image processing and analysis. Kiel University, 1999.

[19] Wareham R, Cameron J, Lasenby J. Applications of conformal geometric algebra in computer vision and graphics. in Computer Algebra and Geometric Algebra with Applications. Berlin, Germany: Springer, 2005: 329-349.

[20] West J B, Fitzpatrick J M, Wang M Y, et al. Comparison and evaluation of retrospective intermodality image registration techniques. in Medical Imaging 1996. International Society for Optics and Photonics, 1996: 332-347.

Chapter 6

GA-STIP: Action Recognition in Multi-channel Videos with Geometric Algebra Based Spatio-temporal Interest Points

6.1 Introduction

In recent years, human action recognition has become an emerging research topic in the computer vision field, which has achieved satisfying accuracy and performed effectively in real applications. However, it is still a challenging task to recognize human actions under scale variance and background interference conditions. Videos that display a certain category of human actions typically requires same or similar features about the spatio-temporal arrangement of relevant activity parts at multiple scales. Without such features these actions could be easily confused, since they generally have very similar photometric and background, at both local and global scales.

It is worth notice that the feature analysis has become one of the most important steps in processing and analysis of videos. Harris detector[1] has been successful used in various image processing tasks to locally detect and describe interest points, such as object recognition, robotic mapping and navigation, image stitching, 3D modeling, image similarity and classification.

In order to represent more complex data, traditional feature analysis methods have been extended to higher-dimensional space[2−4]. A spatio-temporal extension method is capable to detect interest points in both spatial and temporal domains.

The extracted features should be invariant to scale, location and orientation changes. Mikolajczyk and Schmid[5] proposed the Harris-Laplace detector. Based on Harris-Laplace detector and the local invariant characters of gray video, Laptev et al.[6] first introduced a novel Harris 3D spatio-temporal corner detector to detect spatio-temporal interest points (STIPs). STIPs represent points where the local neighborhood has a significant variation in the spatio-temporal domain of the video.

The spatio-temporal regions around these interest points are then described using the 3D SIFT descriptor. The original SIFT descriptoris extended to 3D spatio-temporal architecture and has been used for human action recognition in video sequences[7−14]. The computation of local position-dependent histograms in the 2D SIFT algorithm are extended from two to three dimensional to describe SIFT features. The sub-regions surrounding the interest point are sampled, where each pixel contains a single magnitude value and two orientation values.

These descriptors of STIPs have been widely applied to human action recognition. Based on these descriptors, many works have been proved to be effective[15−24]. Niebles[19] proposed a spatio-temporal bag-of-words (BoW) method which represented videos by a codebook. First, STIPs of video are detected and calculated as descriptors. Thereafter, BoW generates a codebook by clustering these descriptors, and then STIPs are quantified into visual codewords. Finally, support vector machines (SVMs) are used to identify human actions.

In analysis of multi-channel videos, the Harris 3D detector is one of the most useful methods for feature extraction in a plenty of applications. However, based on the traditional methods, the multi-channel video is often converted to a gray video. It causes not only a great loss of video information, which degrades the efficiency of the extracted features, but also failure in capture of the inner-channel information of color video because the algorithm extracts features from each channel independently.

In this chapter, we propose a novel STIP extraction algorithm in GA space for multi-channel videos. Our main contribution is incorporating the GA theory into the traditional Harris 3D algorithm and Gaussian pyramid framework. As the traditional detector algorithms fail to detect features effectively from the multi-channel videos, we utilize the property of GA theory in order to represent a multi-channel video with multi-channel information. Specifically, we propose a new method to obtain the scale space of multi-channel videos by GA representation. Then, we use BoW and SVM to perform classification for action recognition.

6.2 Related Work

6.2.1 Feature Extraction Algorithms Based on Hessian Matrix

To capture spatio-temporal events in video, methods[9,11] transform the 3-dimensional (add time axis to the 2-dimensional space) video signal to spatio-temporal matrix, followed by extraction of interest point. The method is described as follows.

Stage 1 Spatio-temporal Second-moment Matrix

To achieve the invariance in both space and time, a spatio-temporal second-moment matrix is calculated. Interest points are extracted in three different planes—xy, yt and xt. Any random point can carry vital information in both the spatial and the temporal domains.

For a multi-channel video sequence with a $W \times H$ size, let $f(x, y, t)$ denote the pixel at (x, y) in frame t. The convolution of multi-channel video signal f with the 3D Gaussian filter G constructs the linear scale-space representation L:

$$L(x, y, t; \sigma, \tau) = G(x, y, t; \sigma, \tau) f(x, y, t) \tag{6.1}$$

The Gaussian filter G is an extension of the 2D in time dimension.

The spatio-temporal Gaussian function with the spatial and the temporal scale parameters, σ and τ, is given by

$$G(x, y, t; \sigma, \tau) = \frac{1}{(2\pi)^{3/2}\sigma^2\tau} \exp\left(-\frac{x^2+y^2}{2\sigma^2} - \frac{t^2}{2\tau^2}\right) \tag{6.2}$$

To produce a spatio-temporal second-moment matrix with the Gaussian filter at scales σ_i and τ_i, differential operators are used to compute the second-order derivatives of L.

$$\mu(p) = \int_{q\in\mathbb{R}^3} (\nabla L(q))(\nabla L(q))^{\mathrm{T}} g(p-q; \sigma_i, \tau_i)\mathrm{d}q \tag{6.3}$$

where $\nabla L = (L_x, L_y, L_t)^{\mathrm{T}}$.

Stage 2 Interest Points Detection and Description

Once the spatio-temporal matrix μ is constructed, points with significant variations of multi-channel video pixel values can be detected by maximizing all eigenvalues $\lambda_1, \lambda_2, \lambda_3$ of μ. Therefore, local extrema of the interest point in the xy, yt and xt planes can be obtained by carrying out the following operator $H = \det \mu - k(\text{trace}\mu)^3$.

Similar to the original 2D SIFT proposed by Lowe[25], the local extrema is detected by comparing each sample point to its eight neighbor-points in the matrix H. And, filtering is applied to remove noisy points and edges.

Extrema is detected from each block separately, and STIPs which fit the requirement of a detailed defined model can be extracted. In the model, the stable candidate points are marked as STIPs in a specific scale of the spatio-temporal matrix space. These points are described by the vectorized gradient values of the 3D neighborhood surrounding the point. In our method, the 3D SIFT is used to robustly compute the vectorized gradient values.

Based on the spatio-temporal space, extended algorithms with greater performance have been proposed utilizing various detectors and descriptors such as Harris 3D[5,26], Cuboid detector[27], Hessian detector[28], 3D SIFT[12], HOG3D[29]. However, the existing algorithms simply represent a single-channel video or a gray video instead of a multi-channel video. For example, although Harris 3D algorithm[26] use spatio-temporal matrix to obtain the spatio-temporal interest points of gray video, it is not capable of processing multi-channel videos and fail to preserve multiscale information. In addition, many algorithms of detecting or describing the features of color images have been proposed in recent years[30−33]. Li et al.[34] proposed a new scale invariant feature transform for multi-channel image using geometric algebra.

To make the spatio-temporal matrix be able to process the videos in different scales with more channels (i.e. the multi-channel videos), we propose a GA-based spatio-temporal interest points (STIP) detection method.

6.2.2 Geometric Algebra (GA)

William K. Clifford introduced geometric algebra (GA)[35−38], also known as Clifford algebra, which is an efficient tool for geometric representation and computation. GA provides a useful framework together with a set of effective algorithms for expansion, elimination and simplification. Besides, the coordinate-free property of GA enables it to easily carry out the geometric calculation and analysis in higher dimensional space. It has already become an important research tool for theoretical mathematics and physics[39−41].

Let $\mathbb{G}_n$ is n-dimensional geometric algebra with an orthogonal basis $e_1, e_2, \cdots, e_n$. If $a \in \mathbb{G}_n$, then, a can be represented as

$$a_1, a_2, \cdots, a_n \in \mathbb{R}. \tag{6.4}$$

where $a_1, a_2, \cdots, a_n \in \mathbb{R}$.

In general, geometric product of the basis elements is not commutative. Two basis vectors e_i and e_j have many properties. It is denoted as follows.

$$e_i e_j = -e_j e_i, \quad \forall i \neq j \tag{6.5}$$

$$e_i^2 = 1 \tag{6.6}$$

For vectors a and b, the geometric product is defined as

$$ab = a \cdot b + a \wedge b \tag{6.7}$$

where $a \cdot b$ represents the inner product; $a \wedge b$ represents the outer product. The geometric product has the properties as the following.

$$a \cdot b = \frac{1}{2}(ab + ba) \tag{6.8}$$

$$a \wedge b = \frac{1}{2}(ab - ba) \tag{6.9}$$

The GA shows the complete mathematical relationship between the two vectors geometric product which is neither fully symmetric, nor fully anti-symmetric. In GA space, multi-vectors are the basic elements. Accordingly, a multi-vector V in $\mathbb{G}_n$ can be represented by

$$\begin{aligned} V = & E_0(V) + \sum_{1 \leqslant i \leqslant n} E_i(V) e_i + \sum_{1 \leqslant i \leqslant j \leqslant n} E_{ij}(V) e_{ij} \\ & + \cdots + E_{1 \cdots n}(V) e_{1 \cdots n}, E(V) \in \mathbb{R}^{M \times N} \end{aligned} \tag{6.10}$$

6.3 The GA-STIP Algorithm for Multi-channel Video

In this section, we present the process of the GA-STIP in details. As shown in Fig.6.1, the main idea of our algorithm is to incorporate the GA theory into the new scale space of the multi-channel videos, and then detect the STIPs based on the representation of the multi-channel videos in GA.

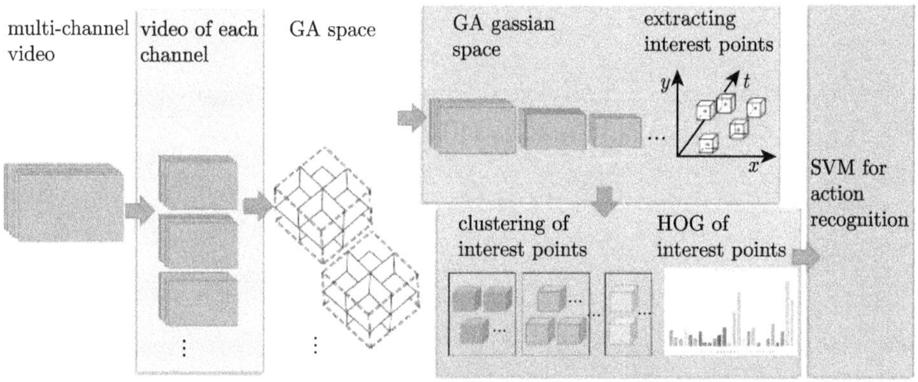

Fig.6.1 Overview flow chart of the GA-STIP approach

6.3.1 Representation of Multi-channel Video in GA

Suppose an n-dimensional space $\mathbb{G}_n$, f is expressed as $f = f_0 + \sum_{i=1}^{n} f_i$ in GA. Similarly, a multi-channel video also can be represented in GA as follows.

$$f(x, y, t) = \sum_{i=1}^{n} f_i(x, y, t) e_i \tag{6.11}$$

where $f(x, y, t)$ represents a video frame of a multi-channel video at t moment, $f_i(x, y, t)$ is the pixel matrix of the i-th-channel of the video frame $f(x, y, t)$, and (x, y, t) is the 3D-coordinate in space-time space.

As a color video is the most common case which needs to be processed in reality, and is also a special case of the multi-channel video with 3 channels, we take color videos as an example to verify the effectiveness of our algorithm. The color video can be represented as following:

$$f(x, y, t) = \sum_{i=1}^{3} f_i(x, y, t) e_i \tag{6.12}$$

where the dimension represents channels of color video, there are several color spaces such as the RGB (i.e. red, green and blue), HSI (i.e. Hue, saturation and intensity), and HSV (i.e. Hue, saturation and value). The most common color space is RGB which is used in this chapter.

6.3.2 Spatio-temporal Interest Points of Multi-channel Video

In the traditional feature point extraction algorithm, they firstly convert the multi-channel video into gray scales, and then carry out convolutional computation between Gaussian filter and gray videos to get the corner points with significant variations. The most important step is to use Gaussian filter to blurred video, which extracts edge and different resolution features. In our proposed method, we use a new Gaussian filter to process multi-channel video which is represented in the GA space.

1. STIP Detection in GA Spatio-temporal Matrix

In multi-channel video, spatial and temporal information is represented by 3-dimensional coordinates (x, y, t). In addition, there is inner-channel information of n-dimensional space. Linking all information together is an n+3 dimensional in GA space. The spatial-temporal GA space is defined as $R^{n+3}_{\text{chan,spat,temp}}$, where n denotes the channel number of multi-channel video $(f_1, f_2, f_3, \cdots, f_n)$, 3 denotes the 3-dimensional coordinate of spatial and temporal (x, y, t). The corresponding GA is referred to as G_{n+3}. As shown in Fig. 6.2, the coordinate of $R^{n+3}_{\text{chan,spat,temp}}$ is represented by $(x, y, t; f_1, f_2, f_3, \cdots, f_n)$.

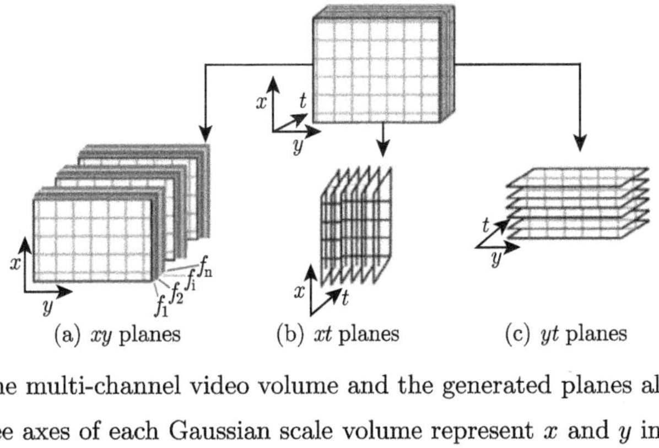

Fig. 6.2 The multi-channel video volume and the generated planes along different directions. Three axes of each Gaussian scale volume represent x and y in the spacial and t in the temporal dimensions, creating three planes, xy, xt and yt

Because of the complexity of multi-channel dimensions, the 2-dimensional Gaussian function is not capable of processing video. Therefore, a novel 3-dimensional

Gaussian function to process multi-channel of video is defined as follows.

$$G_{n+3}(x,y,t;\sigma,\tau)=\frac{1}{(2\pi)^{3/2}\sigma^2\tau}\exp\left(-\frac{x^2+y^2}{2\sigma^2}-\frac{t^2}{2\tau^2}\right)e_i \tag{6.13}$$

where (x,y,t) denotes the 3D coordinate, σ_i and τ_i are the corresponding spatial and temporal scale parameters of Gaussian filter. In experiments, the Gaussian convolution kernel is used in the convolutional computation of the multi-channel video. In $\mathbb{G}_{n+3}$, the Gaussian convolution kernel from the Eq.(6.13) can be described as Fig. 6.3, where $g_{ijk}=\sum_{l=1}^{n} g_{ijkl}e_l$ and $g_{ijkl}\in\mathbb{R}, i,j,k$ denotes the 3-dimensional coordinate of spatial and temporal information, l denotes the corresponding channel.

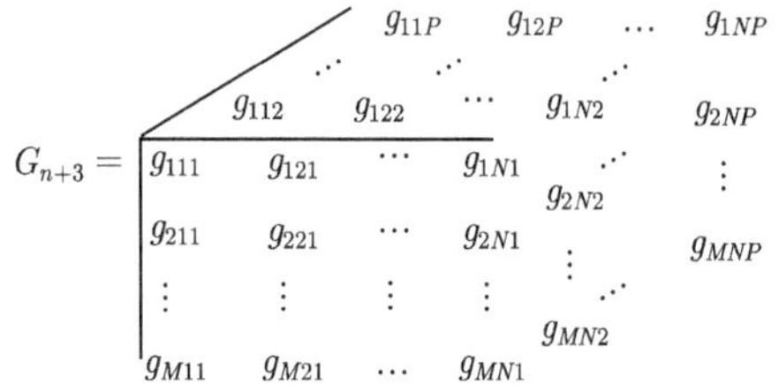

Fig. 6.3 Shows the distribution of each element in space

According to the discussion above, the 1-side convolution of the multi-channel video $f(x,y,t)$ and the Gaussian filter $\mathbb{G}_{n+3}(x,y,t;\sigma_l,\tau_l)$ in $\mathbb{G}_{n+3}$ is

$$\begin{aligned}L_1&=G_{n+3}(x,y,t;\sigma_i,\tau_i)\otimes f(x,y,t)\\&=\frac{1}{MNP}\sum_{i=0}^{M-1}\sum_{j=0}^{N-1}\sum_{k=0}^{P-1}\Big(G_{n+3}(i,j,k;\sigma_i,\tau_i)\,f(x-i,y-j,t-k)\Big)\end{aligned} \tag{6.14}$$

where the product of $\mathbb{G}_{n+3}(x,y,t;\sigma_l,\tau_l)$ and $f(x-i,y-j,t-k)$ is the geometric product.

Based on the 1-side convolution, the 2-side convolution of the multi-channel video $f(x,y,t)$ and the Gaussian filter $\mathbb{G}_{n+3}(x,y,t;\sigma_l,\tau_l)$ in $\mathbb{G}_{n+3}$ which constructs a spatio-temporal scale-space representation L is

$$\begin{aligned}L&=L_1\otimes G_{n+3}(x,y,t;{\sigma'}_l,{\tau'}_l)\\&=\frac{1}{MNP}\sum_{i=0}^{M-1}\sum_{j=0}^{N-1}\sum_{k=0}^{P-1}\Big(G_{n+3}(i,j,k;\sigma_l,\tau_l)\end{aligned}$$

$$\left(f\left(x-i, y-j, t-k\right) \otimes G_{n+3}\left(x, y, t ; \sigma'_l, \tau'_l\right)\right)\Big)$$

$$=\frac{1}{MNP} \sum_{i=0}^{M-1} \sum_{j=0}^{N-1} \sum_{k=0}^{P-1}\left(G_{n+3}\left(i, j, k ; \sigma_l, \tau_l\right)\right.$$

$$\left.\left(\frac{1}{MNP} \sum_{q=0}^{M-1} \sum_{r=0}^{N-1} \sum_{s=0}^{P-1}\left(f\left(x', y', t'\right) G_{n+3}\left(q, r, s ; \sigma'_l, \tau'_l\right)\right)\right)\right)$$

$$=\frac{1}{M^2 N^2 P^2} \sum_{i=0}^{M-1} \sum_{j=0}^{N-1} \sum_{k=0}^{P-1} \sum_{q=0}^{M-1} \sum_{r=0}^{N-1} \sum_{s=0}^{P-1}$$

$$\left(G_{n+3}\left(i, j, k ; \sigma_l, \tau_l\right) f\left(x', y', t'\right) G_{n+3}\left(q, r, s ; \sigma'_l, \tau'_l\right)\right) \tag{6.15}$$

where $x'=x-i-q, y'=y-j-r, t'=t-k-s$. M, N and P indicate the size of the convolution window. The product of $\mathbb{G}_{n+3}(\cdot)$ and $f(\cdot)$ is geometric product.

Then GA is applied to L to get GA spatio-temporal scale-space. Let

$$L_2=G_{n+3}\left(i, j, k ; \sigma_l, \tau_l\right) f\left(x', y', t'\right) G_{n+3}\left(q, r, s ; \sigma_l', \tau_l'\right) \tag{6.16}$$

$$G_{n+3}\left(i', j, k ; \sigma_l, \tau_l\right)=\sum_{i=1}^n g_i\left(i', j, k ; \sigma_l, \tau_l\right) e_i \tag{6.17}$$

$$f\left(x', y', t'\right)=\sum_{i=1}^n f_i\left(x', y', t'\right) e_i \tag{6.18}$$

$$G_{n+3}\left(q, r, s ; \sigma_l', \tau_l'\right)=\sum_{i=1}^n g_i\left(q, r, s ; \sigma_l', \tau_l'\right) e_i \tag{6.19}$$

To reduce the complexity of formula deduction, we use g_i to replace $g_i\left(i', j, k ; \sigma_l, \tau_l\right)$, f_i to replace $f_i\left(x', y', t'\right)$, g_i' to replace $g_i\left(q, r, s ; \sigma_l', \tau_l'\right)$.

Accordingly, L_2 can be expanded as the following

$$\begin{aligned} L_2 &=\left(\sum_{i=1}^n g_i e_i\right)\left(\sum_{i=1}^n f_i e_i\right)\left(\sum_{i=1}^n g_i' e_i\right) \\ &=\left(\sum_{i=1}^n g_i f_i+\sum_{i=1}^n \sum_{j=1}^n g_i f_j\left(e_i \wedge e_j\right)\right)\left(\sum_{i=1}^n g_i' e_i\right) \\ &=\sum_{i=1}^n \sum_{j=1}^n g_i g_j' f_i e_j+\sum_{i=1}^n \sum_{j=1}^n \sum_{k=1}^n g_i g_k' f_j\left(e_i \wedge e_j \wedge e_k\right) \end{aligned} \tag{6.20}$$

So L is transformed in GA spatio-temporal space as follows.

$$L=\mathbb{G}_{n+3}\left(x, y, t ; \sigma_l, \tau_l\right) \otimes f(x, y, t) \otimes G_{n+3}\left(x, y, t ; \sigma_l', \tau_l'\right)$$

$$=\frac{1}{M^2N^2P^2}\sum_{i=0}^{M-1}\sum_{j=0}^{N-1}\sum_{k=0}^{P-1}\sum_{q=0}^{M-1}\sum_{r=0}^{N-1}\sum_{s=0}^{P-1}\left(\sum_{h=1}^{n}\sum_{l=1}^{n}g_hf_hg_l^{'}e_l\right.$$
$$\left.+\sum_{h=1}^{n}\sum_{l=1}^{n}\sum_{u=1}^{n}g_hf_lg_u^{'}\left(e_h\wedge e_l\wedge e_u\right)\right)\Bigg) \tag{6.21}$$

$\sum_{h=1}^{n}\sum_{l=1}^{n}g_hf_hg_l^{\prime}e_l$ has already contained the information of each channel of the multi-channel video, so in experiments, we only use $\sum_{h=1}^{n}\sum_{l=1}^{n}g_hg_l^{\prime}f_he_l$ to deduce the following function as

$$L=G_{n+3}\left(x,y,t;\sigma_l,\tau_l\right)\otimes f\left(x,y,t\right)\otimes G_{n+3}\left(x,y,t;\sigma_l^{\prime},\tau_l^{\prime}\right)$$
$$=\frac{1}{M^2N^2P^2}\sum_{i=0}^{M-1}\sum_{j=0}^{N-1}\sum_{k=0}^{P-1}\sum_{q=0}^{M-1}\sum_{r=0}^{N-1}\sum_{s=0}^{P-1}\left(\sum_{h=1}^{n}\sum_{l=1}^{n}g_hf_hg_l^{'}e_l\right)$$
$$=\sum_{l=1}^{n}\left(\frac{1}{M^2N^2P^2}\sum_{i=0}^{M-1}\sum_{j=0}^{N-1}\sum_{k=0}^{P-1}\sum_{q=0}^{M-1}\sum_{r=0}^{N-1}\sum_{s=0}^{P-1}\sum_{h=1}^{n}g_hf_hg_l^{'}\right)e_l \tag{6.22}$$

For a given scale of observation σ_l and τ_l, STIPs can be found from spatio-temporal second-moment matrix which is composed by the second-order derivatives of L to compute a second-order matrix at each point of the video. The spatio-temporal matrix μ at 3D pixel coordinates $p(x,y,t)$ can be defined as

$$\mu=\mathrm{g}(x,y,t;\sigma_l,\tau_l)\otimes\begin{pmatrix}{L_x}^2 & L_xL_y & L_xL_t\\ L_xL_y & {L_y}^2 & L_yL_t\\ L_xL_t & L_yL_t & {L_t}^2\end{pmatrix}$$
$$=\int_{q\in\mathbb{R}^3}(\nabla L(q))(\nabla L(q))^{\mathrm{T}}\mathrm{g}(x,y,t;\sigma_l,\tau_l)\mathrm{d}q \tag{6.23}$$

where $\nabla L(p)=(L_x,L_y,L_t)^{\mathrm{T}}$.

Then, spatio-temporal Harris corner response function H is designed as Eq.(6.26). The coordinates of the local maximum-valued points, which are also the coordinates of interest points, can be obtained in the H. In order to generate the response function H, the eigenvalues of the spatio-temporal matrix need to be calculated, and then be used to carry out the addition and multiplication operation. The threshold value of the H is set to determine whether the corresponding point is an interest point.

Let

$$\mu = \begin{pmatrix} a_{11} & a_{12} & a_{13} \\ a_{21} & a_{22} & a_{23} \\ a_{31} & a_{32} & a_{33} \end{pmatrix} \tag{6.24}$$

where a_{ij} and 3-dimensional data of the video are consistent in the dimensions. Therefore, the Hadamard product is used to compute the rank and value of matrix.

For the same dimension A and B, the Hadamard product is a matrix, which have the same dimension as the operands. The elements of the Hadamard product given by $(A \cdot B)_{ij} = (A)_{ij}(B)_{ij}$.

$$\begin{aligned} \det(\mu) = & a_{11} \cdot (a_{22} \cdot a_{33} - a_{32} \cdot a_{23}) \\ & - a_{21} \cdot (a_{12} \cdot a_{33} - a_{32} \cdot a_{13}) \\ & - a_{31} \cdot (a_{12} \cdot a_{23} - a_{22} \cdot a_{13}) \end{aligned} \tag{6.25}$$

$$\text{trace}(\mu) = a_{11} \cdot a_{22} \cdot a_{33} \tag{6.26}$$

The response function H is produced by combining the determinant and the trace of μ in the following way. Based on $\det(\mu)$ and $\text{trace}(\mu)$, the response function H can be computed as follows. The value of k is generally set at 0.04~0.06.

$$H = \det \mu - k(\text{trace}(\mu))^3 \tag{6.27}$$

The final result of the function H is a matrix, the size of which is $N \times M \times L$. $N \times M$ represents the size of frames of the video. L denotes the total number of frames of video. As shown in Fig. 6.4, local extrema is detected by comparing each sample point with its eight surrounding neighbor points and the corresponding nine points in both frames former and latter.

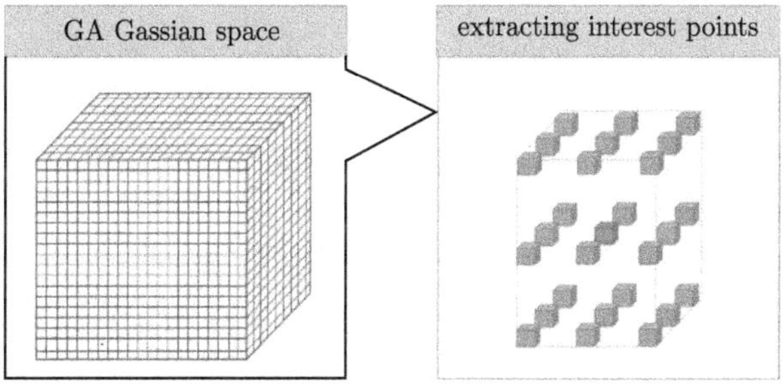

Fig. 6.4 Local extrema is detected by comparing each sample point to its eight neighbor points in the current frame and the nine corresponding neighbor points in both frames former and latter

2. GA Spatio-temporal of Gaussian Pyramid

However, using single Gaussian kernel to extract feature lacks multi-scale information. To achieve the invariance in both space and time, we calculate a spatio-temporal Gaussian pyramid. The principle of the scale space is to imitate the operating principle of human retina, which is similar to get the contour of the object when we observe distant objects. And it means that the larger the scale, the more blurred is the image. Therefore, it is very necessary to establish Gaussian pyramid for feature extraction.

At the same scale, feature values are extracted by different Gaussian variances σ_l and τ_l, The Gaussian scale space is as follows.

$$L = \{L(x, y, t; \sigma_l, \tau_l) \,|\, l = 1, 2, \cdots, m\} \tag{6.28}$$

The parameters are defined as $\sigma_l = (k_1)^l \sigma$, $\tau_l = (k_2)^l \tau$. where l is the parameter to control the number of layers, and generally l is equal to 3.

As shown in Fig.6.5, signal can be downsampled spatially and temporally from the last scale of the former layer and generates a layer with the lower frame rate and smaller-sized frames. The frame size at level G_i is $\frac{W}{2^i} \times \frac{H}{2^i}$.

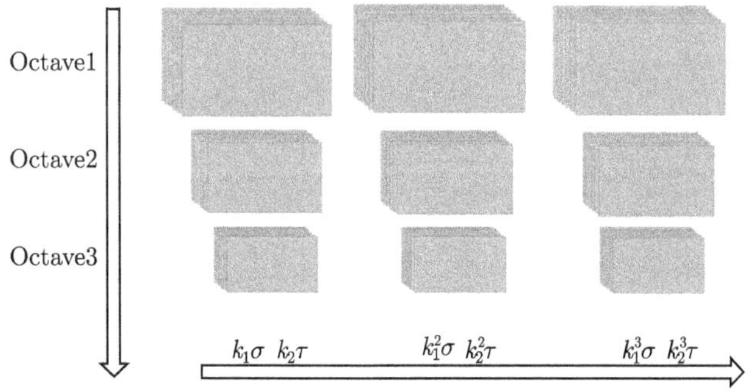

Fig. 6.5 Using different Gaussian cores on the multi-channel video space

The method of scale space derives from the traditional algorithm. At different scales, we can get more stable interest points by merging feature points. The most significant advantage is that our method is capable of detecting feature points on multiple channels when processing multi-channel video. Consequently, when dealing with complex videos, this novel algorithm will make better use of multi-scale and multi-channel information.

6.3.3 Spatio-temporal Descriptors of Feature Points

We assign each STIP one or more orientations based on local gradient directions, which is the key step in achieving invariance to rotation. Given a STIP in the Gaussian video space $L(x, y, t; \sigma_l, \tau_l)$, it can be obtained from formula (6.16) that the local gradient of the STIP is calculated from the i-th-channel of the Gaussian video space. The gradient magnitude $m_{3D}(x, y, t)$ and orientation $\theta(x, y, t)$ can be computed through the pixel of neighbor points:

$$m_{3D}(x, y, t) = \sqrt{{L_x}^2 + {L_y}^2 + {L_t}^2} \tag{6.29}$$

$$\theta(x, y, t) = \arctan(L_y / L_x) \tag{6.30}$$

$$\varphi(x, y, t) = \arctan\left[L_t / ({L_x}^2 + {L_y}^2)^{1/2}\right] \tag{6.31}$$

where $L_x = |L_c(x+1, y, t, \Sigma_\sigma, \Sigma_\tau) - L_c(x-1, y, t, \Sigma_\sigma, \Sigma_\tau)|$, $L_y = |L_c(x, y+1, t, \Sigma_\sigma, \Sigma_\tau) - L_c(x, y-1, t, \Sigma_\sigma, \Sigma_\tau)|$, $L_t = |L_c(x, y, t+1, \Sigma_\sigma, \Sigma_\tau) - L_c(x, y, t-1, \Sigma_\sigma, \Sigma_\tau)|$.

Compared with the traditional 2D SIFT in the calculation of pixel direction, it can be observed that φ now encodes the angle away from the 2D gradient direction. The direction angle θ represents the angle between the projection of the vector on the XY plane and the X axis, and the angle φ represents the angle between the vector and the XY plane. Because of the fact that $L_x^2 + L_y^2$ is positive, φ will always be in the range $\left(-\frac{\pi}{2}, \frac{\pi}{2}\right)$. This is a desired effect, causing every angle to be represented by a single unique (θ, φ) pair.

As is shown in Fig.6.6, the video sequence is divided into spatio-temporal cubes, which are pixel points spaces extracted surrounding the interest points in video. The arrow of each pixel point denotes the gradient direction, and the length means gradient value. The left sphere. The arrow in the left sphere represents the gradient direction, which is denoted by θ and φ. The gradient amplitude and direction of each pixel in the cube is calculated by using formula (6.28) $\sim$ (6.30). Since the area occupied by each bin on the sphere is different, the closer to two poles of the sphere, the smaller the area is occupied by bin, so it should be normalized. Therefore, it makes more sense if the area of each bin is equal.

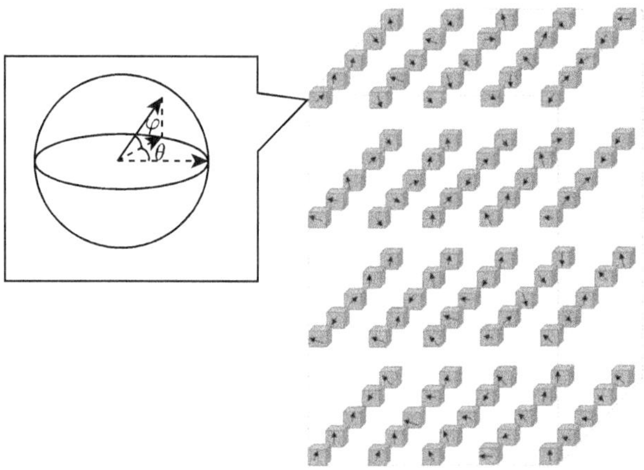

Fig. 6.6 Spherical gradient direction. A unique group (θ and φ) representing the gradient direction

As is shown in Fig.6.7, we divide the sphere approximately as 80 congruent triangles. We define a 3-dimensional vector in each triangle, which points from the center of the sphere to the center of triangle and then forms a 80×3 center vector set $V_{\text{center}} = \{V_{\text{center1}}, V_{\text{center2}}, \cdots, V_{\text{center80}}\}$. Then, the direction of each pixel in the cube is calculated as follows.

$$V_{3\text{D}}(x, y, t) = (L_x, L_y, L_t)/m_{3\text{D}}(x, y, t) \tag{6.32}$$

The next step is to determine the projection distance of the direction vector of each pixel in the cube to V_{center}. The green line in Fig.6.7 denotes the projection distance, and the longest distance is considered as the vector direction of the pixel point. Gradient amplitude of pixel points constructs a weighted histogram for the 3D neighborhood points around a given interest point. When constructing gradient histogram of STIPs, we used bins to represent the given interest point.

6.3.4 Action Recognition of the Multi-channel Video

1. Clustering of Feature Descriptors

To detect similar events in the video data, it is necessary to classify the descriptors first, which benefits action recognition. The same action produces different feature

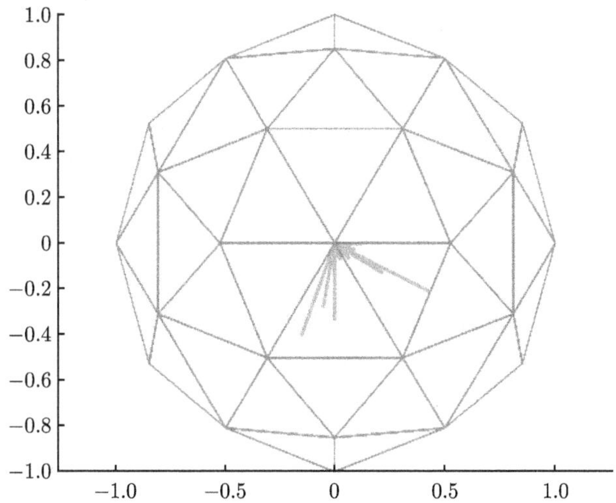

Fig. 6.7 80 gradient directions. The whole gradient histogram contains 80 bins

points in different videos, but feature descriptors of them are inherently similar. After classifying feature descriptors, vectors can be used to represent the internal correlation of each action.

Feature descriptors in this chapter are clustered using K-means clustering algorithm. The feature descriptors of all the videos on the training set are collected into the same data set. We derive n clustering centers, each of which is considered as spatio-temporal vetor $W = \{f_1, f_2, f_3, \cdots, f_m\}$, where m denotes the dimension of the feature descriptor. All spatio-temporal vetors $V = \{w_1, w_2, w_3, \cdots, w_n\}$ of the data set is called spatio-temporal codebook.

2. Classification of Human Action

According to the above steps, we can train spatio-temporal codebook corresponding to different action classes. We describe a description subset as $M = \{m_1, m_2, m_3, \cdots, m_s\}$, where s means the number of descriptors for a multi-channel video.

Based on the spatio-temporal codebook $V = \{w_1, w_2, w_3, \cdots, w_n\}$, we calculate the Euclidean distance of each spatio-temporal vector w_i and the feature descriptor in spatio-temporal codebook. The feature descriptors of each training set video are classified according to the spatio-temporal codebook, and a histogram vector can be acquired. The histogram vector can be expressed as $H = \{h_1, h_2, h_3, \cdots, h_n\}$.

Finally, a nonlinear support vector machine (SVM) is applied to classify. The formula for calculating the kernel function of SVM is as follows.

$$K(H_i, H_j) = \exp\left(-\frac{1}{2A}\sum_{m=1}^{n}\frac{(h_{im} - h_{jm})^2}{h_{im} + h_{jm}}\right) \tag{6.33}$$

where H_i and H_j is histogram vector, n is a space-time codebook dimension. A represents the average distance between all training samples.

Instead of feature descriptors, SVM identifies actions by the histogram vectors of training set.

6.3.5 The Implementation of GA-STIP

We summarize the proposed algorithm (GA-STIP) for human actions recognition by the following steps.

Step 1. Represent an input multi-channel video $f(x, y, t)$ in the GA by Eq. (9.11). Each frame in the multi-channel video is represented by a multi-vector.

Step 2. Remove the luminance part of the frames and retain the chromaticity frames $f_{\text{ch}}(x, y, t)$. It can reduce the influence of the luminance change[34].

Step 3. Use the convolution and Gaussian filter in the GA to obtain the scale space L and GA-Gaussian scale space of $f_{\text{ch}}(x, y, t)$ by Eq. (6.27).

Step 4. Locate local maxima of the matrix H as STIPs under different Gaussian scales.

Step 5. Describe detected feature points using Eq.(6.28)-(6.30) with the GA.

Step 6. Create space-time codebook.

Step 7. Train human action classifier.

Here we take the color video as an example. The color video consists of three channels, so it can be expressed as a formula (6.12). Then we make a 2-side convolution of the color video, and we can get the simplified channel fusion expression Eq.(6.22). The interest points are calculated based on Harris 3D algorithm. By changing Gaussian variances σ_l and τ_t, the Gaussian scale space can be constructed as shown in Fig.6.5. Therefore, we can obtain the interest points at different scales. The histogram of feature points in video can be obtained by 3D SIFT feature descriptors. Then, the BoW clustering algorithm is used to form the spatio-temporal codebook through video feature histogram clustering. Finally, the feature of the video is classified in the spatio-temporal codebook to form a one-dimensional vector, which can be classified by SVM.

6.4 Experimental Analysis

6.4.1 Data Sets

The Weizmann database (http://www.wisdom.weizmann.ac.il/~vision/SpaceTime-Actions.html), UT-interaction (http://cvrc.ece.utexas.edu/SDHA2010/Human_Interaction.html) database and UCF11 database (http://crcv.ucf.edu/data/UCF_YouTube_Action.php) are tested for the classification effect and include 16 different action categories. The Weizmann dataset contains 10 categories, which are bend, jack, jump, pjump, run, side, skip, walk, wave 1 and wave 2.93 action videos have completed by 9 performers. Another UT-interaction dataset contains 6 categories and 120 action videos, which are shaking, hugging, kicking, pointing, punching and pushing.

6.4.2 Experimental Analysis

In this section, we will combine theoretical and experimental results to analyze the method of this chapter, and discuss the advantages of the GA-STIP algorithm in extracting spatio-temporal interest points, and the effect of clustering features on subsequent steps.

1. Comparison of Temporal and Spatial Feature Points

GA-STIP algorithm can extract information from each channel in multi-channel video. Therefore, more effective feature points can be extracted from the feature matrix in GA space. In order to verify the effectiveness of the GA-STIP algorithm, the comparison of the number and the effective of spatio-temporal interest points will be performed under the same threshold conditions. Another kind of comparison algorithm selects Harris 3D to extract spatio-temporal interest points under the corresponding gray video.

Under the same experimental environment set up for two different datasets, the result is showed in Fig.6.8 and Fig.6.9. Fig.6.8 shows the number of STIPs detected in gray video and original video in 10 action categories of Weizmann database. Fig.6.9 shows the number of STIPs detected in gray video and original video in 6 action categories of UT-interaction database. The number of STIPs extracted from original color video is far more than that of STIPs extracted from gray video. There are two reasons for the good effect of GA-STIP algorithm. Firstly, with different spatial and temporal scale parameters applied in feature extracted, multi-scale information can be detected. Compared with Harris 3D calculated using a fixed-scale space, the number of STIPs extracted by our algorithm are more abundant. Secondly, multi-channel video contains more information than single-channel video, so GA-STIP algorithm can detect more STIPs by using original color video directly.

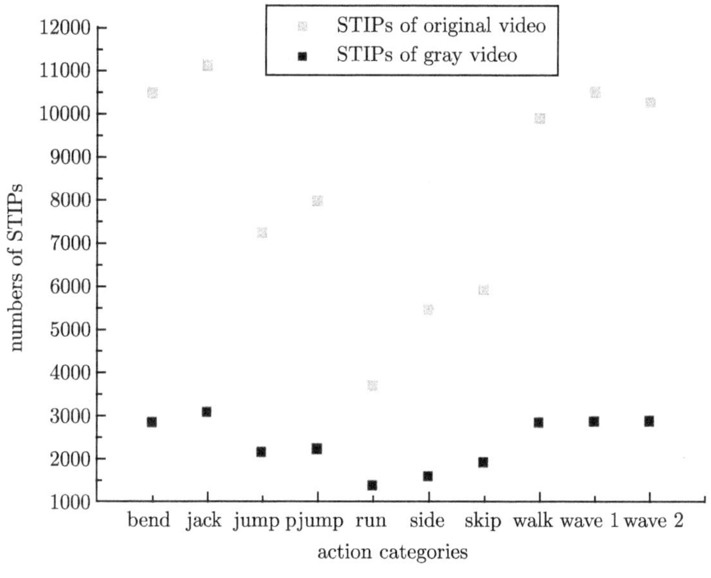

Fig. 6.8 Number of STIPs in Weizmann data sets

However, the detected STIPs cannot all be used, a large number of which located in the background need to be excluded. After the filtering algorithm, the comparison of effective STIPs is as follows.

In Fig.6.10, we choose 4 as both the spatial and temporal scale parameters. (a) shows STIPs of example frames of Weizmann database detected by GA-STIP algorithm (top row) and Harris 3D algorithm (bottom row). (b) shows STIPs of example frames of UT-interaction database detected by GA-STIP algorithm (top row) and Harris 3D algorithm (bottom row).

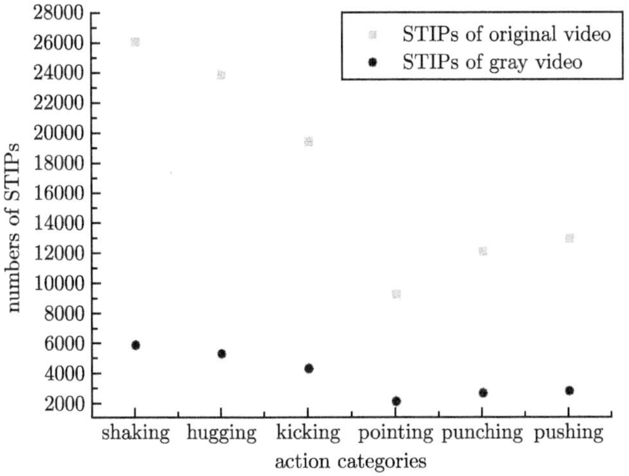

Fig. 6.9 Number of STIPs in UT-interaction data sets

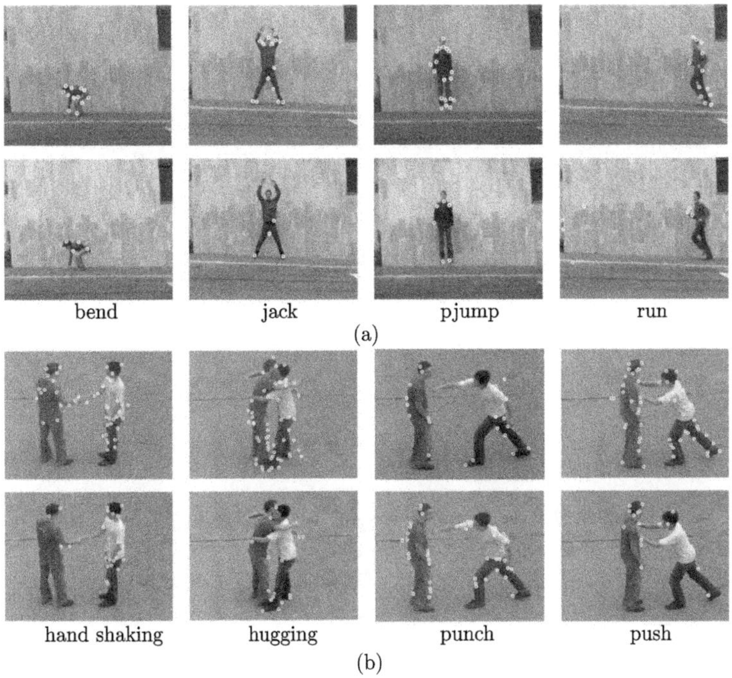

Fig. 6.10 The effective of STIPs in Weizmann and UT-interaction data sets

In the experiment, the extracted STIPs need to be filtered through a filtration algorithm to eliminate the STIPs existing in the background. From the experimental results, it is clear that the STIPs final-extracted by the proposed algorithm basically exclude the false STIPs from background, which contribute to the effectiveness of STIPs that should be basically on the target person. In contrast, STIPs extracted using traditional Harris 3D algorithm to capture movement features of target person turn to be much less after filtered through a filtration algorithm and especially, plenty STIPs in Fig.6.11 (a) (4) still exist on the background.

While traditional Harris 3D algorithm extract STIPs based on gray videos, the proposed GA-STIP algorithm take advantage of the multi-channel information existing in color videos, preserve the RGB inner-relationship and achieve more effective STIPs.

As can be seen from Fig.6.11, our algorithm shows the advantages of extraction of STIPs by using different spatial and temporal scale parameters. In Fig.6.11,

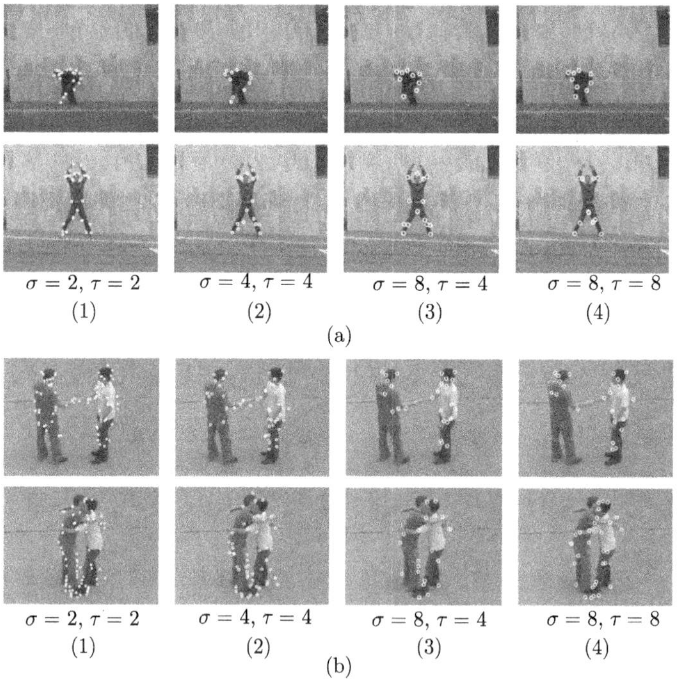

Fig. 6.11 The effect of STIPs using different spatial and temporal scale parameters of σ and τ

(a) represents the number and position of STIPs in Weizmann database; (b) represents the number and position of STIPs in UT-interaction database. Smaller spatial and temporal scale parameters can detect rich local features, while large spatial and temporal scale parameters can detect features dependencies, especially sensitive to inflection points of motion.

The experimental results show that when both spatial and temporal scale parameters are smaller, as $\sigma = 2, \tau = 2$, GA-STIP algorithm extracts more points, while with larger parameters, as $\sigma = 8, \tau = 8$, it puts much more emphasis on the features of bent joints.

Fig.6.12 shows the STIP extracted using UCF11 data set when we choose 2 as both the spatial and temporal scale parameters. The GA-STIP algorithm combines the advantages of both, and the number of STIPs are more abundant and more stable.

Fig. 6.12 The effective feature points extracted in UCF11 data set by our method

2. Advantage Analysis of GA-STIP

From the above analysis, it is obvious that the GA-STIP algorithm can obtain a large number of STIPs in the multi-channel video as compared to the gray video detected using the Harris 3D algorithm. Similarly, the GA-STIP algorithm also shows great advantages in effective spatio-temporal interest points. Consequently, it can be concluded that when performing spatio-temporal interest point clustering, good effects are exhibited due to the number of effective feature points.

6.4.3 Experimental Results

We calculate the histograms of the visual-words, using the extracted local motion features, and concatenate them to a final feature set for SVM learning. Multiple classifiers are used to study the feature set, and generate $N \times \frac{N-1}{2}$ SVM classifier.

1. Action Classification of Weizmann

The Weizmann data set is one of the majority-used databases for human action recognition. We select the videos of the dataset for training and testing. The leave-one-out method is used to compute recognition accuracy. During the training process of the Weizmann dataset, the k-means clustering algorithm was used to cluster the sample feature set, and the space-time codebook of the sample space has been established. The selection of clustering dimension has an impact on the recognition performance. After experimental observation, the dimension of spatio-temporal codebook is 1000.

Table 6.1 shows confusion matrix of Weizmann data set, and we can observe the action recognition effect of ST-STIP method on Weizmann data set. Table 6.2 shows the effect of action recognition on Weizmann data set compared with the Harris 3D method. The results suggest that our system can identify better. In Table 6.3, we compared our system with the previous systems on detecting actions from Weizmann data set.

Table 6.1 Confusion matrices when classifying human action of Weizmann

	bend	jack	jump	pjump	run	side	skip	walk	wavel	Wave2
bend	1.00	0	0	0	0	0	0	0	0	0
jack	0	1.00	0	0	0	0	0	0	0	0
jump	0	0	1.00	0	0	0	0	0	0	0
pjump	0	0	0	1.00	0	0	0	0	0	0
run	0	0	0.10	0	0.90	0	0	0	0	0
side	0	0	0.11	0	0	0.89	0	0	0	0
skip	0	0	0.10	0.10	0	0	0.80	0	0	0
walk	0	0	0	0	0	0	0	1.00	0	0
wavel	0	0	0	0	0	0	0	0	1.00	0
wave2	0	0	0	0	0	0	0	0	0.11	0.89

Table 6.2 Comparison with the Harris 3D method

method	recognition accuracy/%
our model	94.8
Harris 3D	88.6

Table 6.3 Comparison of features extracted by different algorithms using gray videos

method	recognition accuracy/%
our Model	94.8
MOSIFT with SP[2]	93.7
MOSIFT[2]	91.9
ROI[3]	92.1
ST features[4]	84.3
3D SIFT[12]	82.6

2. Action Classification of UT-Interaction

The UT-interaction dataset is also a very popular dataset about the daily interaction of human beings. These high-level actions contain a lot of complex information, which involves the combination of discriminatory atomic motions and human interactions. The UT-interaction is divided into two sub datasets, each of which contains 6 types of actions and 10 videos for each class. The leave-one-out method is used to compute recognition accuracy on each subset of data, and finally the performance has been averaged. When we process the UT-interaction dataset, the method is the same as that of Weizmann data set. After experimental observation, the dimension of spatio-temporal codebook is 5000.

Table 6.4 shows the action recognition effect of ST-STIP method on UT-interaction dataset and compared with the Harris 3D methods. It can be observed that our method can achieve better recognition accuracy.

3. Action Classification of UCF11

UCF11 data set is very challenging due to large variations in camera motion, object appearance and pose, object scale, viewpoint, cluttered background, illumination conditions, etc. In our experiments, we select some videos of which the camera

Table 6.4 Performance comparison on the UT-interaction dataset

interaction	recognition accuracy by our model	recognition accuracy by Harris 3D
shake hands	0.85	0.70
hug	0.90	0.90
point	0.90	0.85
punch	0.70	0.60
kick	0.85	0.65
push	0.55	0.40
total	0.792	0.683

does not mobile. These videos are divided into 6 categories, which are basketball shooting, diving, golf swinging, soccer juggling, tennis swinging, and walking with a dog. The leave-one-out method is used to compute recognition accuracy on each subset of data, and at last the performance is averaged. While the dimension of spatio-temporal codebook in the traditional Harris 3D algorithm is 1800, in our method, the dimension of spatio-temporal codebook is 3500.

Table 6.5 shows that the comparison between our method and traditional Harris 3D algorithm, which verifies that the performance of our algorithm is better than other algorithms in each category.

Table 6.5 Performance comparison on the UCF11 dataset

action	recognition accuracy by our model	recognition accuracy by Harris 3D
basketball shooting	0.80	0.70
diving	0.90	0.80
golf swinging	0.70	0.50
soccer juggling	0.80	0.60
tennis swinging	0.70	0.60
walking with a dog	0.40	0.30
total	0.717	0.583

4. Discussion

In the above experiments, we find that the recognition accuracy is quite low for the category "walking with a dog" in the UCF11 data set. It is because that the video footage we select for this category keep moving and cause some points in the background to be extracted as feature points. In the process of background point noise processing, if the threshold value set by the filtering algorithm is too low, it could not effectively filter the noise points. In contrast, when the threshold value is too high, it could not extract enough useful STIP, which lowers the final recognition rate. This is a defect inherent in the traditional Harris 3D algorithm, which can only extract effective feature points from camera-fixed video. In the chapter, our main contribution is the improvement in the traditional Harris 3D algorithm for the fixed camera video. This part of performance improvement can be reflected in our comparative experiments. This chapter is the first to apply geometric algebra in video processing, and the fusion of video multi-channel information can be realized by the nature of geometric algebra. This idea can also be used in the latest algorithms, such as machine learning algorithms. Currently, in the field of machine learning, geometric algebra has been used in deep learning algorithm for image processing. Therefore, geometric algebra can be combined with deep learning algorithm for action recognition of video, which is believed to have better effects.

6.5 Conclusions

In this chapter, we propose a novel algorithm based on the GA theory. Considering that the GA representation of the multi-channel videos contains rich channel information, we combine the spatio-temporal matrix with GA space to detect and describe the local STIPs in multi-channel videos. In order to complete the detection and location of the feature points in the multi-scale spaces, GA Gaussian filters are proposed and applied to obtain the scale space of multi-channel videos. We utilize the spatio-temporal gradient vectors to describe the STIPs in the GA. compared with results of previous detection algorithms, our proposed approach leads to more robust feature points found in all channel videos of a multi-channel video. By evaluating the experimental results with action dataset, we have confirmed that the GA-STIP has improved performance in extraction of features from a multi-channel video and can recognize complicated human activities. The proposed GA-STIP achieves more accurate results in analysis of multi-channel videos.

At next steps, we will combine geometric algebra with convolutional neural network for action recognition on complex scene.

References

[1] Harris C, Stephens M J. A combined corner and edge detector. New York: Proc Alvey Vision Conf, 1988: 147-151.

[2] Bebars A, Hemayed E E. Comparative Study for Feature Detectors in Human Activity Recognition. 2013 9th International Computer Engineering Conference (ICENCO), 2013: 19-24. 10.1109/ICENCO.2013.6736470.

[3] Khedher M I, Yacoubi M E, Dorizzi B. Human action recognition using continuous HMMs and HOG/HOF silhouette representation. in International Conference on Pattern Recignittion and Methods, Vilamoura, Portugal, 2012: 503-508.

[4] Liu J, Ali S, Shah M. Recognizing Human Actions Using Multiple Features. 2008 IEEE Conference on Computer Vision and Pattern Recognition, 2008: 1-8. 10.1109/CVPR.2008.4587527.

[5] Mikolajczyk K, Schmid C. Indexing Based on Scale Invariant Interest Points. Proceedings Eighth IEEE International Conference on Computer Vision. ICCV 2001, Vancouver, BC, Canada, 2001, 1: 525-531. 10.1109/ICCV.2001.937561.

[6] Laptev I, Lindeberg T. Space-time Interest Points. in: 9th International Conference on Computer Vision, 2003: 432-439.

[7] Flitton G. Object Recognition using 3D SIFT in Complex CT Volumes. Proceedings of the British Machine Vision Conference, 2010, 1: 1-12.

[8] Ullah M M, Parzi S N, Laptev I. Improving bag-of-features action recognition with non-local cues. BMVC, 2010, 10: 1-95.

[9] Laptev I. On space-time interest points. International Journal of Computer Vision, 2005: 107-123.

[10] Laptev I, Caputo B, Schüldt C, et al. Local velocity-adapted motion events for spatio-temporal recognition. Computer Vision and Image Understanding, 2007: 207-229. 10.1016/j.cviu.2006.11.023.

[11] Lopes A P B, Oliveira R S, de Almeida J M, et al. Spatio-Temporal Frames in a Bag-of-Visual-Features Approach for Human Actions Recognition. Xxii Brazilian Symposium on Computer Graphics and Image Processing IEEE Computer Society, 2009: 315-321.

[12] Scovanner P, Ali S, Shah M. A 3-dimensional Sift Descriptor and Its Application to Action Recognition. Proceedings of the 15th International Conference on Multimedia,

2007: 357-360. 10.1145/1291233.1291311.

[13] Chen M, Hauptmann A. MoSIFT: Recognizing Human Actions in Surveillance Videos. Annals of Pharmacotherapy, 2009: 150-152.

[14] Al Ghamdi M, Zhang L, Gotoh Y. Spatio-temporal SIFT and Its Application to Human Action Classification. International Conference on Computer Vision Springer-Verlag, 2012: 301-310.

[15] Ryoo M, Aggarwal J. UT-Interaction Dataset, ICPR Contest on Semantic Description of Human Activities (SDHA), 2010.

[16] Ryoo M S, Aggarwal J K. Spatio-temporal relationship match: Video structure comparison for recognition of complex human activities. Proceedings, 2009: 1593-1600.

[17] Ryoo M S. Human Activity Prediction: Early Recognition of Ongoing Activities from Streaming Videos. 2011 International Conference on Computer Vision, Barcelona, 2011: 1036-1043. 10.1109/ICCV.2011.6126349.

[18] Dollar P, Rabaud V, Cottrell G, et al. Behavior recognition via sparse spatio-temporal features. 2005 IEEE International Workshop on Visual Surveillance and Performance Evaluation of Tracking and Surveillance, 2005: 65-72. 10.1109/VSPETS.2005.1570899.

[19] Niebles J C, Wang H, Li F F. Unsupervised learning of human action categories using spatial-temporal words. International Journal of Computer Vision, 2008: 299-318.

[20] Shen M, Wang R, Cao W. Joint sparse representation model for multi-channel image based on reduced geometric algebra. IEEE Access, 10.1109/ACCESS.2018.2819691.

[21] Raptis M, Sigal L. Poselet Key-Framing: A Model for Human Activity Recognition. 2013 IEEE Conference on Computer Vision and Pattern Recognition, Portland, OR, 2013: 2650-2657. 10.1109/CVPR.2013.342.

[22] Yao A, Gall J, Van Gool L. A Hough Transform-based Voting Framework for Action Recognition. IEEE Conference on Computer Vision and Pattern Recognition, 2010: 2061-2068. 10.1109/CVPR.2010.5539883.

[23] Vahdat A, Gao B, Ranjbar M, et al. A Discriminative Key Pose Sequence Model for Recognizing Human Interactions. 2011 IEEE International Conference on Computer Vision Workshops (ICCV Workshops), Barcelona, 2011: 1729-1736. 10.1109/ICCVW.2011.6130458.

[24] Zhang Y, Liu X, Chang M, et al. Spatio-temporal phrases for activity recognition. ECCV'12 Proceedings of the 12th European Conference on Computer Vision, 2012: 707-721.

[25] Lowe D G. Object Recognition from Local Scale-invariant Features. Proceedings of the International Conference on Computer Vision, 1999: 1150-1157. 10.1109/ICCV.

1999.790410.

[26] Laptev I, Lindeberg T. Local Descriptors for Spatio-temporal Recognition. Spatial Coherence for Visual Motion Analysis. Berlin Heidelberg: Springer, 2006: 91-103. 10.1007/11676959_8.

[27] Dollár P, Rabaud V, Cottrell G, et al. Behavior Recognition via Sparse Spatio-temporal Features. Visual Surveillance and Performance Evaluation of Tracking and Surveillance, 2005. 2nd Joint IEEE International Workshop on, IEEE, 2005: 65-72.

[28] Willems G, Tuytelaars T, Van Gool L. An efficient dense and scale-invariant spatio-temporal interest point detector. European Conference on Computer Vision Springer-Verlag, 2008: 650-663.

[29] Klaser A, Marszałek M, Schmid C. A Spatio-temporal Descriptor Based on 3D-gradients. in: BMVC 2008-19th British Machine Vision Conference, British Machine Vision Association, 2008: 275-281.

[30] Abdel-Hakim A E, Farag A A. CSIFT: A SIFT Descriptor with Color Invariant Characteristics. in: IEEE Conference on Computer Vision and Pattern Recognition, 2006: 1978-1983.

[31] Bosch A, Zisserman A, Munoz X. Scene classification using a hybrid generative/ discriminative approach. IEEE Trans. Pattern Anal. Mach. Intell., 2008: 712-727.

[32] Ai D N, Han X, Ruan X, et al. Color independent components based SIFT descriptors for object/scene classification. IEICE Trans. Inf. & Syst., 2010: 2577-2586.

[33] Li C, Ma L. A new framework for feature descriptor based on SIFT. Pattern Recogn. Lett., 2009: 544-557.

[34] Li Y S, Liu W M, Li X T, et al. GA-SIFT: A new scale invariant feature transform for multispectral image using geometric algebra. Information Sciences, 2014: 559-572.

[35] Buchholz S, Sommer G. On Clifford neurons and Clifford multi-layer perceptrons. Neural Networks the Official Journal of the International Neural Network Society, 2008: 925-935.

[36] Chapman A, Vishne U. Clifford algebras of binary homogeneous forms. Journal of Algebra, 2012: 94-111.

[37] Hestenes D, Sobczyk G. Clifford algebra to geometric calculus, a unified language for mathematics and physics. American Journal of Physics, 1984: 510, 511.

[38] Sommer G. Geometric Computing with Clifford Algebras. Heidelberg: Springer, 1998.

[39] Li Y. A digital image watermarking algorithm based on clifford algebra. Acta Electronica Sinica, 2008: 852-855.

[40] Cao W, Lyu F, He Z, et al. Multimodal medical image registration based on feature

spheres in geometric algebra. in IEEE Access, 2018, 6: 21164-21172. 10.1109/ACCESS.2018.2818403.

[41] Shen M, Wang R, Cao W. Joint sparse representation model for multi-channel image based on reduced geometric algebra. IEEE Access, 10.1109/ACCESS.2018.2819691.

Chapter 7

GA-CNNs: Convolutional Neural Networks Based on Geometric Algebra

7.1 Introduction

Recently, convolutional neural networks (CNNs)[1,2] have been widely applied in varieties of computer vision tasks, such as object detection[3,4], face detection[5,6] or image classification[7,8]. Many researches tend to focus on promoting CNNs models[9,10] and training algorithms to address higher dimensional signal processing problems.

It is worthy reviewing the development of neural networks in different algebraic systems. Aizenberg et al.[11,12] proposed complex-valued neural networks with multi-valued neurons, and it showed the remarkable properties of multi-valued neurons which are used as basics units in neural networks. Clarke[13] developed a new complex-valued neural network which introduced activation function to complex domain in the complex-valued neural networks. Since then, the back-propagation algorithm for complex neural network was proposed by Leung[14], Benvenuto[15], Nitta[16] and Georgiou[17], this algorithm provides a method for updating the complex-valued neural network and has been successfully applied in classification of complex signals. Haensch et al.[18] firstly applied complex-valued convolutional neural network to object detection in polarimetric synthetic aperture radar image, although the architecture of this network is only one convolution layer, the experiment results are superior to those of complex-valued multilayer perceptrons. However, complex-valued neural networks are restricted by higher dimensional signals. Nitta[19,20] extended neural network to quaternion domain, furthermore, a version of the back-propagation algorithm based on quaternion was also provided. Kominami[21] introduced multi-valued convolutional neural network and verified the improved learning ability on handwritten digital image dataset. Nevertheless, signals in real world are complicated and multi-dimensional, existing CNNs models still face some challenges to process multi-dimensional signals.

Fortunately, GA has ability to address the problems mentioned above. GA serves as a unified generation of real-valued, complex-valued, and many algebraic systems. GA can be applied to many fields such as virtual reality, signal processing and computer vision[22−26]. In this chapter, we propose convolutional neural networks based on Geometric Algebra (GA-CNNs) which is capable of processing multi-dimensional signals with fewer neurons.

Our contribution is fully utilizing the advantages of GA to represent different geometric entities to build GA-CNNs model which has the ability to avoid processing multi-dimensional signals independently. Rather than real-valued CNNs that regard each channel as a real number, which inevitably ignore the interrelationship among different channels, our proposed GA-CNNs model successfully preserve the inherent structure by representing the multi-dimensional signals as GA multivectors. Therefore, GA-CNNs model can capture inherent structures in multi-dimensional signals. The classification experiments show that the proposed model outperforms real-valued CNNs with respect to higher classification accuracy and faster convergence.

The rest of this chapter is organized as follows: Section 7.2 introduces the basics of geometric algebra and neural network based on geometric algebra. Section 7.3 describes the proposed GA-CNNs model in detail. Comparison experiments between GA-CNNs and real-valued CNNs are provided in Section 7.4. Section 7.5 concludes the chapter briefly.

7.2 Related Work

7.2.1 Basics of Geometric Algebra

Geometric algebra (GA)[27−29] is described by William K. Clifford, also called Clifford algebra. For multi-dimensional signals, GA is an efficient framework to handle the representation and computation issues and a useful tool to apply to mathematics and physics[30].

Mathematically, suppose G_n is denoted as a n-dimensional vector space, the basis of $\mathbb{G}_n$ are denoted as

$$\{e_A = e_{a_1} \cdots e_{a_r} \,|A = (a_1, \cdots, a_r) \in B\} \tag{7.1}$$

where $\mathbb{G}_n$ can be denoted as $\mathbb{G}_{p,q}$ and $n = p + q$.

In general, the geometric algebra is non-commutative, the multiplication of geometric algebra will follow the next rules.

$$\begin{cases} e_{a_i}^2 = 1, & a_i = 1, \cdots, p \\ e_{a_i}^2 = -1, & a_i = p+1, \cdots, p+q \\ e_{a_i} e_{a_j} = -e_{a_j} e_{a_i}, & a_i \neq a_j \end{cases} \tag{7.2}$$

The power set Γ of $\{1, \cdots, n\}$ can turn the basis into an ordered one with the index set B

$$\{|B = (a_1, \cdots, a_r) \in \Gamma, 1 \leqslant a_1 \cdots < a_r \leqslant n\} \tag{7.3}$$

For example, the basis in 2^3 vector space can be denoted as

$$e_\varnothing, e_1, e_2, e_3, e_{12}, e_{13}, e_{23}, e_{123} \tag{7.4}$$

For convenience, in the rest of the chapter, $A = a_1 \cdots a_r$ will be denoted as $A = a_{1 \cdots r}$, and $e_\varnothing = e_0$.

The set $\mathbb{G}_n^t$ is denoted as following, representing t-vector part of $\mathbb{G}_n$.

$$G_n^t = \{e_A \,|e_A \in B, |A| = t\} \tag{7.5}$$

An arbitrary element of the geometric algebra can be denoted as the following form.

$$x = \sum_{t=0}^{n} \langle x \rangle_t = \sum_{A \in B} x_A e_A, \quad x_A \in \mathbb{R} \tag{7.6}$$

For example, an element in 2^3 vector space can be denoted as

$$\begin{aligned} x &= \langle x \rangle_0 + \langle x \rangle_1 + \langle x \rangle_2 + \langle x \rangle_3 \\ &= x_0 + x_1 e_1 + x_2 e_2 + x_3 e_3 \\ &\quad + x_{12} e_{12} + x_{13} e_{13} + x_{23} e_{23} + x_{123} e_{123} \end{aligned} \tag{7.7}$$

The addition of the geometric algebra can be defined as

$$x + y = \sum_{A \in B} (x_A + y_A) e_A \tag{7.8}$$

The geometric product of geometric algebra can be defined as

$$x \otimes_{p,q} y = x \cdot y + x \wedge y \tag{7.9}$$

where $x \cdot y$ and $x \wedge y$ represents scalar, bivector part of the geometric product, respectively.

The reversion of geometric algebra can be defined as

$$\left\langle x^{\dagger} \right\rangle_t = (-1)^{\frac{t(t-1)}{2}} \langle x \rangle_t \tag{7.10}$$

The conjugation of geometric algebra can be defined as

$$\langle x \rangle_t^{*} = (-1)^{\frac{t(t+1)}{2}} \langle x \rangle_t \tag{7.11}$$

The module of geometric algebra can be defined as

$$|\langle x \rangle| = \sqrt{\sum_{t=0}^{n} |\langle x \rangle_t|^2} \tag{7.12}$$

$$|\langle x \rangle_t| = \sqrt{\langle x \rangle_t \cdot \langle x \rangle_t} \tag{7.13}$$

7.2.2 Neural Networks Based on Geometric Algebra

GA neurons are first introduced in this section as follows. GA neurons are the basis units for computing GA - CNNs, a GA neuron can be viewed as the mapping function from $(\mathbb{G}_{p,q})^n$ to $(\mathbb{G}_{p,q})$. Although GA neurons are designed for GA - CNNs, it can be applied to real - valued CNNs as well.

$$y = \sum_{i=1}^{n} w_i \otimes_{p,q} x_i + \theta \tag{7.14}$$

where $x_i \in \mathbb{G}_n$, $w_i \in \mathbb{G}_n$, $\theta \in \mathbb{G}_n$ denote the inputs, weights and bias, respectively. In addition, geometric product replaces real multiplication for GA neurons.

Specially, a basic geometric algebra neuron (BGAN) is a mapping function from $\mathbb{G}_{p,q}$ to $\mathbb{G}_{p,q}$.

$$y = w_i \otimes_{p,q} x_i + \theta \tag{7.15}$$

A BGAN has only one multivector input and weight, compared with a real neuron, it will study only one multi-dimensional input entity. The error function is:

$$E_{p,q} = \frac{1}{2}\|d - y\| = \frac{1}{2}\|\text{error}\| \tag{7.16}$$

where $\|\|$ represents the standard Euclidean norm, and d represents the desired output.

$\text{BGAN}_{0,2}$ stands for a dilatation-rotation of 3D Euclidean space. This weight is updated denoted as:

$$w\,(t+1) = w\,(t) - \eta \cdot \frac{\partial E_{0,2}}{\partial w} \tag{7.17}$$

where η, $0 < \eta < 1$ represents the learning rate.

$$\begin{aligned}\frac{\partial E_{0,2}}{\partial w} &= \frac{\partial E_{0,2}}{\partial w_0}e_0 - \frac{\partial E_{0,2}}{\partial w_1}e_1 - \frac{\partial E_{0,2}}{\partial w_2}e_2 - \frac{\partial E_{0,2}}{\partial w_{12}}e_{12} \\ &= \text{error} \otimes_{0,2} x^* \end{aligned} \tag{7.18}$$

Therefore, the amount of the update of weight is the multiplication between the error made by the neuron and the conjugate of its input.

7.3 Convolutional Neural Networks Based on Geometric Algebra (GA-CNNs)

This section describes our proposed convolutional neural networks based on geometric algebra (GA-CNNs) which is built as extension from real domain to GA space. The overall structure of our network is shown in Fig.7.1. In GA-CNNs model, inputs, outputs and weights are GA multivectors. The operations in each layer and training algorithm will be introduced in detail later.

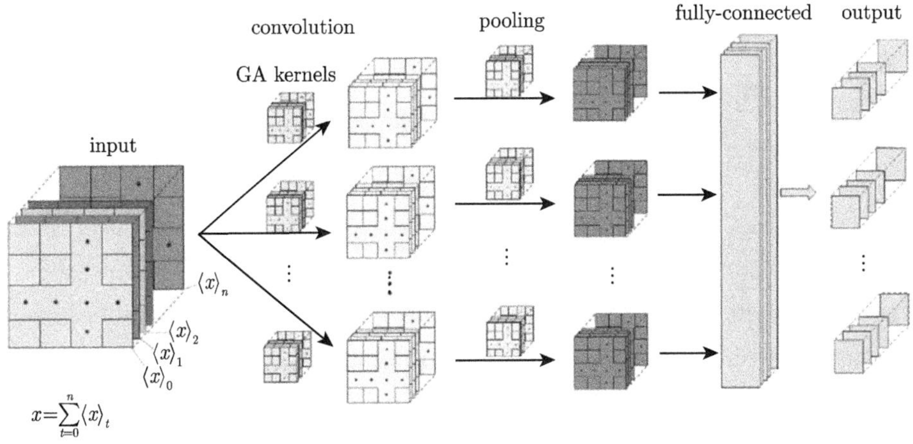

Fig. 7.1 The framework of GA-CNNs

7.3.1 Convolutional Layer

Based on GA, both kernel and input patch are represented as GA multivectors. Convolution with multiple kernels produces multiple channels features for multi-dimensional signals. Compare with real-valued CNNs, it is clear that a convolution based on GA has multiple kernels, as seen in Fig.7.2.

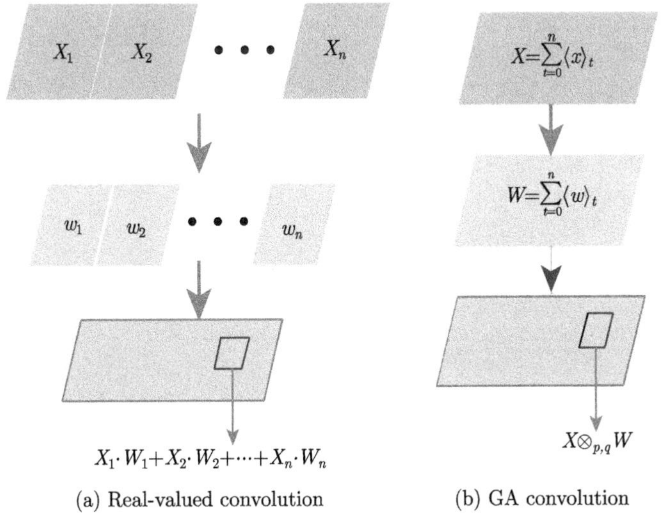

Fig. 7.2 GA and real-valued convolution layers

The output of the convolution layer is $y^{l_{\text{conv}}}$

$$y^{l_{\text{conv}}} = \sum_{A\in B} g^{(l)}\left(\sum_{k}\left(x_k^{(l-1)}\otimes_{p,q} w_{kj}^{(l)} + \theta_{kj}^{(l)}\right)_A\right) \tag{7.19}$$

$$g^{(l)}(x) = \sum_{A\in B} g_R(x_A)\, e_A, \quad x \in \mathbb{G}_n \tag{7.20}$$

where $x_k^{(l-1)}$ represents k-dimensional signals and $w_{kj}^{(l)}$ represents convolution kernels. Similar to real-valued activation function, the sigmoid activation function $g_R(\ )$ is used for all neurons in GA-CNNs model, for the GA multivctors, each of parts goes through the activation function, separately.

7.3.2 Pooling Layer

We propose maximum pooling based on GA as an extension of maximum pooling in the real-valued CNNs.

$$\text{soft max}_\alpha\left(\{x_i\}_{i=1}^n\right) = \sum_{A\in B} \frac{\sum_i x_{iA} e^{\alpha x_{iA}}}{\sum_j e^{\alpha x_{iA}}} e_A, \quad x_{iA} \in \mathbb{R} \tag{7.21}$$

In detail, we obtain with three different cases:

$$\text{soft max}_\alpha\left(\{x_i\}_{i=1}^n\right) = \sum_{A\in B} \frac{\sum_i x_{iA} e^{\alpha x_{iA}}}{\sum_j e^{\alpha x_{iA}}} e_A = \begin{cases} \sum_{A\in B} (\max_i x_{iA})\, e_A, & \alpha \to \infty \\ \sum_{A\in B} \left(\frac{1}{n}\sum_i x_{iA}\right) e_A, & \alpha \to 0 \\ \sum_{A\in B} (\min_i x_{iA})\, e_A, & \alpha \to -\infty \end{cases} \tag{7.22}$$

where $\forall\alpha \in \mathbb{R}, \{x_i\}_{i=1}^n \in (\mathbb{G}_{p,q})^n$.

7.3.3 Fully-connected Layer

As mentioned above, the previous layers extracting more interrelationship features than real-valued CNNs. In order to keep the interrelationship information, we propose the multilayer perceptron based on geometric algebra (GA-MLP) used as the output of the fully-connected layers connecting to all neurons in the last pooling layer. The definition of a GA-MLP is given as follows.

$$s_j^{(L)} = \sum_k w_{kj} \otimes_{p,q} x_k^{(L-1)} + \theta_j^{(L)} \tag{7.23}$$

$$y_j^{(L)} = \sum_{A \in B} g_R \left(\left[s_j^{(L)} \right]_A \right) e_A = \sum_{A \in B} g_R \left(\sum_k \left(w_{kj} \otimes_{p,q} x_k^{(L-1)} + \theta_j^{(L)} \right)_A \right) e_A \tag{7.24}$$

where L represents output layer $L > l$.

7.3.4 Backpropagation Algorithm

This chapter proposes a GA version backpropagation algorithm which can be applied to convolution neural network whose weights, inputs and outputs are all GA. The GA based backpropagation algorithm compute a steepest gradient descent on an error function E with chain rule for training the network. For GA-CNNs, the chain rule is given below.

The standard output square error is

$$E = \frac{1}{2} \sum_{j=1}^{m} \sum_{A \in B} \left([d_j]_A - \left[y_j^{(L)} \right]_A \right)^2 \tag{7.25}$$

where $y_j \in \mathbb{G}_n$ represents the output and $d_j \in \mathbb{G}_n$ represent the desired output in the output layer L.

In the process of backpropagation, the gradient of loss function for weight is

$$\Delta w_{kj}^{(L)} = \sum_{A \in B} \frac{\partial E}{\partial \left[w_{kj}^{(L)} \right]_A} e_A = \sum_{C \in B} \frac{\partial E}{\partial \left[s_j^{(L)} \right]_C} \frac{\partial \left[s_j^{(L)} \right]_C}{\partial \left[w_{kj}^{(L)} \right]_A} \tag{7.26}$$

According to the chain rule, we obtain ${\delta_j}^{(L)}$ as follows.

$$\begin{aligned} \delta_j^{(L)} &= \sum_{C \in B} \frac{\partial E}{\partial \left[s_j^{(L)} \right]_C} e_C \\ &= \sum_{C \in B} \frac{\partial E}{\partial [y_j]_C} \frac{\partial \left[y_j^{(L)} \right]_C}{\partial \left[s_j^{(L)} \right]_C} e_C n \\ &= \left([d_j]_C - \left[y_j^{(L)} \right]_C \right) \left(g^{(L)} \right)' \left(\left[s_j^{(L)} \right]_C \right) e_C \end{aligned} \tag{7.27}$$

Then updating of the weights in fully-connected layer L is

$$\begin{aligned}\Delta w_{kj}^{(L)} &= \sum_{A\in B}\frac{\partial E}{\partial\left[w_{kj}^{(L)}\right]_A}e_A\\ &= \sum_{A\in B}\left(\sum_{C\in B}\left[\delta_j^{(L)}\right]\frac{\partial\left[s_j^{(L)}\right]_C}{\partial\left[w_{kj}^{(L)}\right]_A}\right)e_A\\ &= \left(d_j-y_j^{(L)}\right)\odot\left(g^{(L)}\right)'\left(s_j^{(L)}\right)\otimes_{p,q}\left(x_k^{(L-1)}\right)^*\end{aligned} \tag{7.28}$$

$$\Delta\theta_j^{(L)} = \delta_j^{(L)} \tag{7.29}$$

For two multiple vectors $W = \sum_{A\in B}[w]_A e_A = w_0e_0 + w_1e_1 + \cdots + w_ne_n$ and $Z = \sum_{A\in B}[z]_A e_A = z_0e_0 + z_1e_1 + \cdots + z_ne_n$ with the same dimensions, we define $W\odot Z = w_0z_0e_0 + w_1z_1e_1 + \cdots + w_nz_ne_n$.

For updating the weight in the fully-connected layers $l(l < L)$.

$$\delta_j^{(l)} = \left(\sum_j\left(w_{kj}^{(l+1)}\right)^*\otimes_{p,q}\delta_j^{(l+1)}\right)\odot\left(g^{(l)}\right)'\left(\left[s_j^{(l)}\right]_C\right)e_C$$

$$\begin{aligned}\Delta w_{kj}^{(l)} &= \delta_j^{(l)}\otimes_{p,q}\left(x_k^{(L-1)}\right)^*\\ &= \left(\sum_j\left(w_{kj}^{(l+1)}\right)\otimes_{p,q}\delta_j^{(l+1)}\right)\\ &\quad\odot\left(g^{(l)}\right)'\left(\left[s_j^{(l)}\right]\right)\otimes_{p,q}\left(x_k^{(L-1)}\right)^*\end{aligned} \tag{7.30}$$

Similar with the bias

$$\Delta\theta_j^{(l)} = \delta_j^{(l)} \tag{7.31}$$

The error of pooling layer can be derived as follows.

$$\delta_j^{(l)} = \left(g^{(l)}\right)'\left(s^{(l)}\right)\mathrm{conv2}_G\left(\delta_j^{(l+1)},\mathrm{rot180}\left(w_j^{(l+1)}\right),\text{'full'}\right) \tag{7.32}$$

$$\Delta\,\theta_j^{(l)} = \sum_j\delta_j^{(l)} \tag{7.33}$$

where $\mathrm{conv2}_G(\ )$represents the convolution of multivector matrix, rot180( ) represents an operator which aims to make the convolution function perform cross-correlation with rotating the kernel, 'full' represents an operator which pads the missing inputs with zeros automatically.

The error of convolution layer can be derived as follows

$$\delta^{(l)} = \text{rot}180\left(w_j^{(l)}\right)^* \otimes_{p,q} \delta^{(l+1)} \odot \left(g^{(l)}\right)' \left(\left[s_j^{(l)}\right]_C\right) e_C \tag{7.34}$$

where rot180( ) represents the rotation of jth the convolution kernel $w_j^{(l)}$ in the convolution layer.

Similarly, the updating weights in the convolutional layer is derived as follows.

$$\begin{aligned}\Delta w_{kj}^{(l+1)} &= \delta_j^{(l)} \otimes_{p,q} \left(x_k^{(l-1)}\right)^* \\ &= \left(\sum_j \left(w_{kj}^{(l)}\right) \otimes_{p,q} \delta_j^{(l-1)}\right) \\ &\quad \odot \left(g^{(l)}\right)' \left(\left[s_j^{(l)}\right]\right) \otimes_{p,q} \left(x_k^{(l-1)}\right)^*\end{aligned} \tag{7.35}$$

7.4 Experiments and Analysis

In this section, the effectiveness of proposed GA-CNNs model is evaluated on three different datasets, compared with real-valued CNNs.

7.4.1 Experiment on Synthetic Data

The first experiment is carried out based on synthetic data. The following three functions have already been used for comparing performances of the GA-CNNs and real-valued CNNs. The network has four layers with a convolution layer with 9 kernels of size 3×3 with a stride of 1 pixel, a pooling layer with size 2×2 and a fully-connected layer. Each function generates 4000 points in [0, 1] and we select 3000 points in each class for training and remaining 1000 points in each class for testing. The training ends at iteration 4500 with epoch 10 and the learning rate was 0.01.

$$\begin{aligned}f_G(x, y) = &(\cos x_1 + \sin y_1) + (\cos x_2 + \sin y_2) e_1 \\ &+ (\cos x_3 + \sin y_3) e_2 + (\cos x_4 + \sin y_4) e_{12}\end{aligned} \tag{7.36}$$

$$h_G(x, y) = \left((x_1 + y_1)^2\right) + \left((x_2 + y_2)^2\right) e_1 + \left((x_3 + y_3)^2\right) e_2 + \left((x_4 + y_4)^2\right) e_{12} \tag{7.37}$$

$$q_G(x, y) = (\exp(-x_1) + \exp(-y_1)) + (\exp(-x_2) + \exp(-y_2)) e_1$$

$$+ (\exp(-x_3) + \exp(-y_3))\, e_2 + (\exp(-x_4) + \exp(-y_4))\, e_{12} \qquad (7.38)$$

To evaluate the performance of the GA-CNNs model with optimal number of convolutional kernels, we consider different numbers of convolutional kernels as 3,6 and 9 to explore the influence of the number of convolutional kernels for the proposed model. In general, the more the number of the convolutional kernels means that the network can obtain more feature maps, it is important for the higher accuracy of classification. As shown in Fig.7.3, it is significant for the GA-CNNs model which converge quickly and achieve the lowest training loss with the same iterations when the number of convolutional kernels is 9. Thus, in this experiment, the number of convolutional kernels is set as 9.

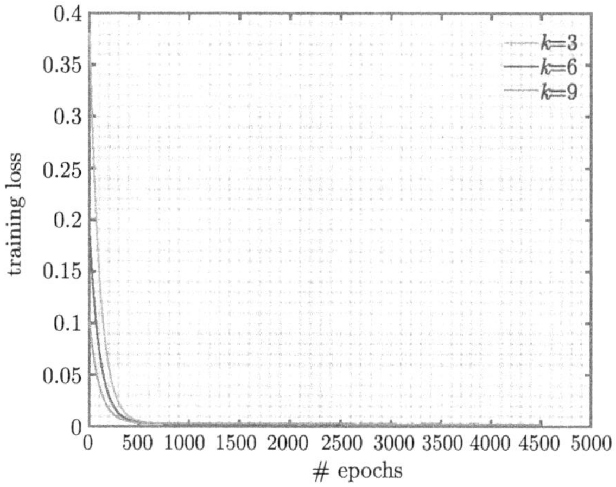

Fig. 7.3 The training loss on GA-CNNs with different number of convolutional kernels

As depicted in Fig.7.4, we can find that the loss function of GA-CNNs converges faster to reach a stable value than real-valued CNNs, that is to say, the proposed GA-CNNs model outperforms the real-valued CNNs model on synthetic data, achieving a smaller value quickly with lower training errors.

For classification accuracy compared between the proposed GA-CNNs model and real-valued CNNs model, the results are listed in Table 7.2. As shown in Table 7.2, the proposed GA-CNNs exhibits higher test accuracy than real-valued CNNs, the proposed GA-CNNs model is 78.7667% while that of real-valued CNNs is 55.5333%. More precisely, the GA-CNNs model has the ability in capturing more information of different channels on synthetic data when representing as GA multivectors. The results have demonstrated that superior performance of GA-CNNs model is achieved compared with real-valued CNNs model on synthetic data.

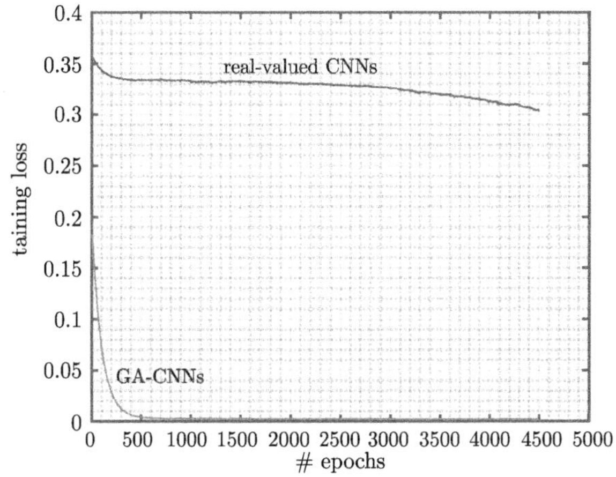

Fig. 7.4 The loss of the real-valued CNNs and GA-CNNs model during training on synthetic data

7.4.2 Experiment on Color Images

The second dataset was captured by a Kaggle dataset, including two sets of images (dogs and cats), The images were cropped into the same small patches of size 32×32 and we select 9000 images for training and 3000 additional images to evaluate our network. The more details can be found at https://www.kaggle.com/c/dogs-vs-cats/data.

For each color image, the real-valued CNNs model deals with three channels independently while the proposed GA-CNNs model processes the inputs with three channels simultaneously, representing the color image as GA multivectors. The network is composed of a convolution layer with 3 kernels of size 3×3 with stride of 1 pixel, a pooling layer with size 5×5 and a fully-connected layer. The training stops at iteration 4500 with epoch 10 and the learning rate was 0.01.

We now select the option number of pooling sizes to evaluate the effectiveness of the proposed GA-CNNs model. The network was trained with different pooling size, including 3×3, 5×5, 10×10, as depicted in Fig. 7.5, when the pooling size is 10×10.

Since the too large pooling size may loss the more position information of feature maps, the pooling size is 10×10 of the network get the slowest convergence and the highest training loss. When the pooling size is 5×5, the proposed GA-CNNs model has fastest convergence and the lowest training loss. Thus, in this experiment, the pooling size is set as 5×5.

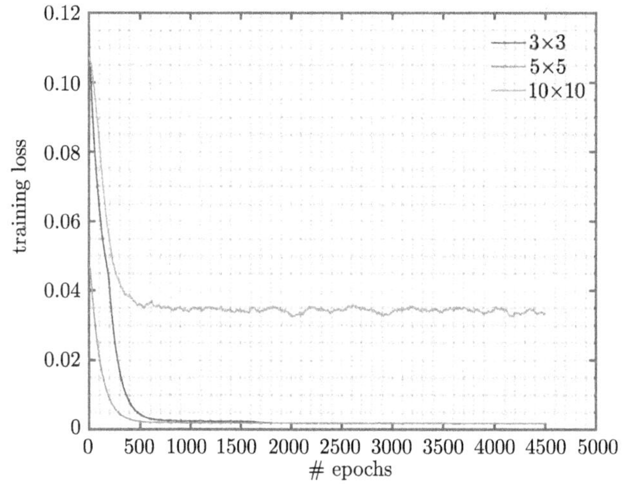

Fig. 7.5 The training loss on GA-CNNs with different pooling size

The results are shown in Fig.7.6, it is clear that GA-CNNs show superior performance than real-valued CNNs on color images dataset. The proposed GA-CNNs model has faster convergence than real-valued CNNs model which means that our model can reach the smaller loss with less iterations than real-valued CNNs model.

As shown in Table 7.2, the GA-CNNs model can also obtain higher accuracy than real-valued CNNs on color image dataset. The proposed GA-CNNs model is 79.8% while that of real-valued CNNs is 64.5%. Rather than real-valued CNNs which process three channels of color image independently, the proposed GA-CNNs model can represent the input as GA multivectors and deal with three channels of color image in a unified way, preserving more representative information among different channels for color images, this information is important for classification. The results have shown that the proposed GA-CNNs model is superior than real-valued CNNs model on color images dataset.

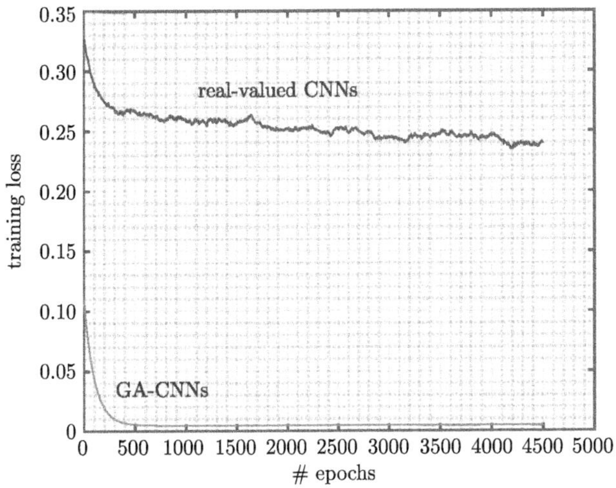

Fig. 7.6 The loss of the real-valued CNNs and GA-CNNs model during training on color image

7.4.3 Experiment on Hyperspectral Images

In the third experiment, the university of Pavia hyperspectral dataset (Pavia U) (Kunkel et al., 1988) are used to validate our model. The dataset consists of 9 classes which cover an urban environment, such as trees, asphalts, shadows, bricks and meadows, etc. as shown in Table 7.1. We just took the 200 images for each class as train data, and 100 images for each class as test data. The available images are composed of 610×340 pixels with 103 spectral bands, spectral range from 0.43 to 0.86μm and its spatial resolution is 1.3m/pixel. The GA-CNNs model processes the 103 spectral bands of each pixel with GA, treating four channels as a unified way. As a result, for GA-CNNs model, each pixel includes 25 spectral bands. The Fig.7.7 shows in the hyperspectral image, there are about 20.62% of pixels which belonging to ground-truth information.

The network for hyperspectral image dataset has one convolution layer with 6 kernels of size 1×4 with a stride of 1 pixel, one pooling layer with pooling size set as 1×2 and one fully-connected layers. In the experiment, the real-valued CNNs and GA-CNNs models have the same structure. The GA-CNNs models are optimized with learning rate set at 0.4 while real-valued CNNs models are optimized with learning rate set at 0.2 and the training ends at iteration 7021 with epoch 20.

Table 7.1 Nine classes of Pavia U

class			samples
No.	color	name	numbers
1		asphalt	6631
2		meadows	18649
3		gravel	2099
4		trees	3064
5		painted metal sheets	1345
6		bare soil	5029
7		bitumen	1330
8		self-blocking bricks	3682
9		shadows	947
total samples			207400

Fig. 7.7 Original ground-truth image of Pavia U scene

In Fig. 7.9, the classification maps were obtained from Pavia U dataset, and in Fig.7.8, we obtain the better convergence of the proposed GA-CNNs model than that of real-valued CNNs model. This result demonstrates that our model can achieve the better performance than real-valued CNNs on hyperspectral images dataset.

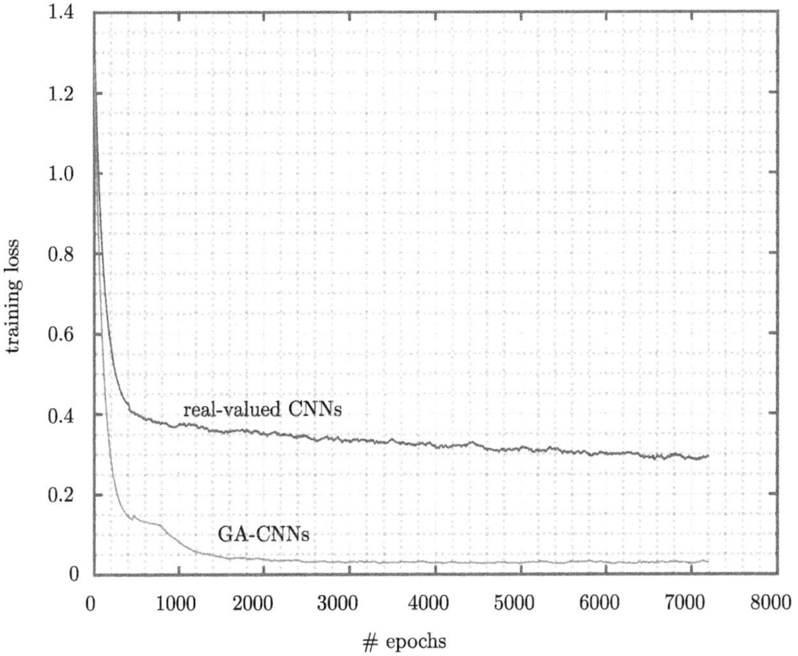

Fig. 7.8 The loss of the real-valued CNNs and GA-CNNs model during training on hyperspectral image

(a) Classification results by real-valued GA-CNNs with background

(b) Classification results by real-valued without background

(c) Classification results by real-valued CNNs model with background

(d) Classification results by GA-CNNs model with background

Fig. 7.9 Classification results for Pavia U dataset

The classification accuracy of the GA-CNNs model is also better than real-valued CNNs on each class for hyperspectral images, as listed in Table 7.2 and Table 7.3. Firstly, GA-CNNs represent hyperspectral images as GA multivectors, preserving more spectral information of different spectral bands. Secondly, our GA-CNNs deal with the GA multivectors with multi-channels, not just sum up the results of each channel finally. So, real-valued CNNs model may loss more representative spectral information for hyperspectral images. Obviously, this experiment further demonstrates that the proposed model is better than real-valued CNNs on hyperspectral images.

Table 7.2 Classification accuracy compared between the proposed GA-CNNs and real-valued CNNs

model	dataset	test accuracy/%
real-valued CNNs	synthetic data	55.5333
	color image	64.5
	hyperspectral image	61.48
GA-CNNs	synthetic data	78.7667
	color image	79.8
	hyperspectral image	70.89

Table 7.3 Classification with spectral-spatial features on the Pavia U dataset

No.	real-valued CNNs/%	GA-CNNs/%
1	74.45	59.86
2	73.02	72.15
3	28.54	73.03
4	52.28	97.78
5	99.26	99.41
6	4.30	41.90
7	9.80	96.02
8	85.63	73.38
9	87.22	100

7.5 Conclusions

In this chapter, the model of convolutional neural networks based on geometric algebra (GA-CNNs) has been proposed as an extension of CNNs from real domain to GA space to accommodate multi-dimensional signal processing. Under the framework of GA, the multi-dimensional signals are significantly regarded as GA multivectors to avoid the loss of interrelationship among channels. Generally, the architecture of the proposed GA-CNNs has been described in detail, including convolution, pooling and fully-connected layers. And we derived the back-propagation training algorithm, in which the weights are represented and updated in the GA space. To evaluate the wide flexibility of our model, we conduct classification experiments on synthetic data, color image dataset and hyperspectral image dataset. The experimental results have demonstrated that the proposed GA-CNNs model possesses faster convergence and higher accuracy, and shows its superiority with compared with current real-valued CNNs models. Moreover, the proposed GA-CNNs model shows its potential as a powerful tool in color image and hyperspectral image processing domain, and more applications such as denoising, detection, fusion, etc., will be explored in our next work.

References

[1] Lang K J, Hinton G E. A time delay neural network architecture for speech recognition. Technical Report CMU-CS-88-152, CMU, 1988.

[2] LeCun Y, Boser B, Denker J S, et al. Backpropagation applied to handwritten zip code recognition. Neural Computation, 1989, 1(4):541–551.

[3] Agrawal P, Girshick R, Malik J. Analyzing the Performance of Multilayer Neural Networks for Object Recognition[C]// European Conference on Computer Vision. Springer International Publishing, 2014:329-344.

[4] Li Y, He K, Sun J, et al. R-FCN: Object detection via region-based fully convolutional networks. In: Advances in Neural Information Processing Systems, 2016: 379-387.

[5] Taigman Y, Yang M, Ranzato M, et al. DeepFace: Closing the Gap to Human-Level Performance in Face Verification[C]// IEEE Conference on Computer Vision and Pattern Recognition. IEEE Computer Society, 2014:1701-1708.

[6] Qin H, Yan J, Li X, et al. Joint Training of Cascaded CNN for Face Detection[C]// Computer Vision and Pattern Recognition. IEEE, 2016:3456-3465.

[7] Zheng Z, Zhang Y, Li L, et al. Classification Based on Deep Convolutional Neural Networks with Hyperspectral Image[C]// Geoscience and Remote Sensing Symposium. IEEE, 2017:1828-1831.

[8] Kulkarni P, Zepeda J, Jurie F, et al. Hybrid Multi-layer Deep CNN/Aggregator Feature for Image Classification[C]// IEEE International Conference on Acoustics, Speech and Signal Processing. IEEE, 2015:1379-1383.

[9] Krizhevsky A, Sutskever I, Hinton G E. Image Net Classification with Deep Convolutional Neural Networks[C]// International Conference on Neural Information Processing Systems. Curran Associates Inc., 2012:1097-1105.

[10] Simonyan K, Zisserman A. Very deep convolutional networks for large-scale image recognition. arXiv preprint arXiv:1409.1556, 2014.

[11] Aizenberg I. Complex-valued Neural Networks with Multi-valued Neurons. Berlin, Heidelberg: Springer, 2011.

[12] Hirose A. Complex-valued Neural Networks. Berlin, Heidelberg: Springer, 2004.

[13] Clarke T L. Generalization of Neural Networks to the Complex Plane[C]// IJCNN International Joint Conference on Neural Networks. IEEE, 1990.

[14] Leung H, Haykin S. The Complex Backpropagation Algorithm. IEEE Press, 1991.

[15] Benvenuto N, Piazza F. On the complex backpropagation algorithm. IEEE Transactions on Signal Processing, 1992, 40(4):967-969.

[16] Nitta T. An extension of the back-propagation algorithm to complex numbers. Neural Networks, 1997, 10(8):1391-1415.

[17] Georgiou G M, Koutsougeras C. Complex domain backpropagation. IEEE Transactions on Circuits & Systems II Analog & Digital Signal Processing, 1992, 39(5):330-334.

[18] Haensch R, Hellwich O. Complex-Valued Convolutional Neural Networks for Object Detection in PolSAR Data[C]// European Conference on Synthetic Aperture Radar. VDE, 2011.

[19] Arena P, Fortuna L, Muscato G, et al. Multilayer perceptrons to approximate complex valued functions. International Journal of Neural Systems, 1995, 6(04):12.

[20] Nitta T. A Quaternary Version of the Back-propagation Algorithm[C]// IEEE International Conference on Neural Networks. IEEE, 1995.

[21] Nitta T. An Extension of the Back-propagation Algorithm to Three Dimensions by Vector Product[C]// International Conference on Tools with Artificial Intelligence. IEEE Xplore, 1993.

[22] Pearson J K, Bisset D L. Neural Networks in the Clifford Domain[C]// IEEE World Congress on IEEE International Conference on Neural Networks. IEEE, 2007.

[23] Kominami Y, Ogawa H, Murase K. Convolutional Neural Networks with Multi-valued Neurons[C]// International Joint Conference on Neural Networks. IEEE, 2017.

[24] Mao X J, Shen C, Yang Y B. Image restoration using very deep convolutional encoder-decoder networks with symmetric skip connections [J]. 2016.

[25] Shang F, Hirose A. Quaternion neural-network-based PolSAR land classification in poincare-sphere-parameter space. IEEE Transactions on Geoscience & Remote Sensing, 2014, 52(9):5693-5703.

[26] Xu Y, Yu L, Xu H, et al. Vector sparse representation of color image using quaternion matrix analysis. Image Processing IEEE Transactions on, 2015, 24(4): 1315-1329.

[27] Buchholz S, Sommer G. On Clifford neurons and clifford multi-layer perceptrons. Neural Netw., 2008, 21(7): 925–935.

[28] Hestenes D, Sobczyk G. Clifford algebra to geometric calculus: A unified language for mathematics and physics. Amer. J. Phys., 1984, 53: 510–511.

[29] Sommer G. Geometric Computing with Clifford Algebras. Berlin, Germany: Springer, 1998.

[30] Shen M, Wang R, Cao W. Joint sparse representation model for multi-channel image based on reduced geometric algebra. IEEE Access, 2018, 6: 24213–24223, doi: 10.1109/ACCESS.2018.2819691.

Chapter 8

Joint Sparse Representation Model for Multi-channel Image Based on Reduced Geometric Algebra

8.1 Introduction

Sparse representations of color image based on dictionary learning have currently been an active area for its potential of providing extremely high performance for various applications[1,2]. In this framework, a color image can be well-represented as a sparse linear combination of elements from an appropriately chosen over-complete dictionary instead of being separated into independent components[3,4]. The quality of dictionary determines the performance of the sparse coding, which can be chosen as a pre-defined set of bases, such as geometric transformations, wavelets[5], curvelets[6], contourlets[7], short-time Fourier kernels[8], Laplacian pyramid[9], or unions of bases.

Traditional monochrome models regard each color image pixel as a scalar and process the pixel in an independent way, which fails to exploit the inter-relationship among the RGB color channels, potentially resulting in some color distortions and blurring effects always appearing in reconstruction results[10,11]. Although the scalar product is modified to enforce average colors, which improves the efficiency of the representation for denoising[1], inpainting[12], classification[13] and object detection[14], some unsatisfying results still persist. For improvement, Mairal et al.[15] proposed a concatenate model by performing a simple concatenation of the RGB values to a single vector and training on those directly to alleviate the hue distortion problem. Unfortunately, the lack of explicit constraints on the correlations among color channels inevitably leads to some unsatisfying results. Another strategy is using independent color channels rather than RGB channels. Pang et al.[16] presented a novel method to perform image colorization using sparse representation in YUV color space, which leads to high-quality colorizations with small number of given color pixels. But due to its limitations, it can only be applied in device dependent image processing, such as demosaicing[17].

Quaternions extend complex numbers with a real part and three imaginary parts[18,19]. Quaternion-based methods have shown the efficiency in dealing with color images[20,21]. Recently, quaternions have been introduced to represent three-channel RGB structures for vector sparse representations of color images[3,22,23], which have shown improved performance for reconstruction[3], denoising[22], inpainting[23] and super-resolution[4]. Xu et al.[23] proposed a vector sparse representation model for color images using quaternion matrix analysis, which conducts the sparse basis selection in quaternion space to successfully avoid the hue bias issue. However, the quaternion-based methods are known to suffer from high computational complexity due to non-commutative multiplication. To improve the method, reduced quaternion matrix (RQM) with commutative multiplication is introduced to reduce computational complexity[24,25]. Gai et al.[26] provided a new vector-valued sparse representation model for color images using RQM.

In general, the existing sparse models treat each color image pixel either as a scalar[1,12–14], which loses some color structures, or as a quaternion vector matrix[22,23,27] with high computational complexity. In this chapter, we propose a novel sparse representation model for multi-channel image based on geometric algebra[28]. In particular, a novel theory of reduced geometric algebra (RGA) is provided with commutative multiplication rules. Taking advantage of the RGA theory, the model represents each color image as a multivector[29] with the spatial and spectral information in RGA space, where the basic operational rules, properties and SVD analysis of color image are defined. We use multivector operations among the color atoms of the learned RGA-based dictionary and the sparse RGA-based coefficients to reconstruct color image blocks. Then, the corresponding dictionary learning method namely K-RGASVD (generalized K-means clustering for singular value decomposition using RGA), is provided. The RGA-OMP (orthogonal matching pursuit in RGA form) method is used to compute the sparse coefficients. Thus it achieves the removal of the redundancy among color channels, the inherent color structures preserved, and low computational complexity.

The rest of this chapter is organized as follows. Section 8.2 reviews the related work. Section 8.3 describes the basic concepts of reduced geometric algebra. The novel RGA-based sparse representation model is provided in Section 8.4. The RGA-based dictionary training algorithm is represented in Section 8.5. In Section 8.6, the reconstruction and denoising experiments are implemented to show the effectiveness and rationality of this algorithm. Finally, Section 8.7 concludes the chapter.

8.2 Related Work

8.2.1 Sparse Representation Models for Color Image

Current researches on image sparse representation proposedz mostly focused on patch processing[1,2]. Denote a color image F, we superimpose all the pixels of each overlapped image patch in each channel and denote the resulting scalar vector as $\boldsymbol{f}_{\mathrm{C}} \in \mathbb{R}^n$, where n equals to the size of the patch vector, the subscript C=R, G, B treats the RGB color channels respectively.

Different dictionaries corresponding to each color channels independently, which proposed by color image sparse models, can be denoted as

$$\boldsymbol{f}_{\mathrm{C}}=\boldsymbol{D}_{\mathrm{C}}\boldsymbol{g}_{\mathrm{C}}, \quad \mathrm{C}=\mathrm{R},\mathrm{G},\mathrm{B} \tag{8.1}$$

where $\boldsymbol{D}_{\mathrm{C}}$ is the corresponding dictionary, and $\boldsymbol{g}_{\mathrm{C}}$ is the assumed sparse coefficient vector. In particular, this monochrome sparse model has achieved state-of-the-art performance for gray-scale i.e. single-channel image processing. However, in terms of color images, it naively represents RGB channels as three independent "gray-scale" images and processes them in a monochrome manner. Owing to the loss of orthogonal property and the lack of inter-correlation among the RGB color channels of this monochrome strategy, it is bound to produce some unsatisfying results, such as hue distortions in the reconstruction results.

In fact, a color image with RGB color channels cannot be just regarded as a scalar, a vector-valued function is employed naturally to denote the structure of color information. Then, the vector form is defined by the pure quaternion form as $\dot{\boldsymbol{f}} = 0 + \boldsymbol{f}_r i + \boldsymbol{f}_g j + \boldsymbol{f}_b k, \dot{\boldsymbol{f}} \in \mathbb{Q}^n$ applying the quaternion representation, the quaternion-based sparse representation model has been given in Ref. [23] as

$$\min_{\dot{\boldsymbol{a}}} \|\dot{\boldsymbol{a}}\|_0, \text{s.t.} \dot{\boldsymbol{f}} = \dot{\boldsymbol{D}}\dot{\boldsymbol{a}} \tag{8.2}$$

where $\dot{\boldsymbol{D}} = \boldsymbol{D}_s + \boldsymbol{D}_r i + \boldsymbol{D}_g j + \boldsymbol{D}_b k, \dot{\boldsymbol{D}} \in \mathbb{Q}^{n \times K}$ is a quaternion dictionary, $\dot{\boldsymbol{a}} = \boldsymbol{a}_0 + \boldsymbol{a}_1 i + \boldsymbol{a}_2 j + \boldsymbol{a}_3 k, \dot{\boldsymbol{a}} \in \mathbb{Q}^K$ is a sparse quaternion coefficient vector corresponding, with its components $\boldsymbol{a}_0, \boldsymbol{a}_1, \boldsymbol{a}_2$ and $\boldsymbol{a}_3 \in \mathbb{R}^K$. Here $\|\dot{\boldsymbol{a}}\|_0$ means the amount of nonzero components in the quaternion coefficient vector.

Compared with the existing monochromatic-based models, it only displays the transformation from gray-scale images to color images, while the quaternion-based method deals with multi-channel information in a unified way to imitate the process of human visual perception. The coefficient matrix captures both the relationship among color channels and the orthogonal property successfully. And the four channel dictionaries uniformly transform $\boldsymbol{f}_r, \boldsymbol{f}_g$ and $\boldsymbol{f}_b$ into a better color space that is orthogonal. The obvious advantage of the quaternion-based method is that it maintains the inherent structures of RGB channels during image reconstruction.

As a matter of fact, the color image is represented as a pure quaternion. However, a series of operations will produce a quaternion containing the real part. Quaternion multiplication consists of sixteen real multiplication and twelve real additions. In addition, the quaternion multiplication is non-commutative, which is not conducive to the design of fast algorithm. Thus, quaternion-based model leads to a large number of data redundancy and high algorithm complexity.

8.2.2 The Basic of Geometric Algebra

William K. Clifford introduced geometric algebra (GA), also known as clifford algebra, which provides a coordinate-free framework to make the computations efficiently[28−31]. It completes the constructions and modellings in a coordinate-free manner and also has revealed wide applications in both physics and computer vision[32−34].

Suppose $\mathbb{G}_n$ is n-dimensional geometric algebra with an orthogonal basis $e_1, e_2, \cdots$,

e_n. If $x \in \mathbb{G}_n$, then, x can be represented as

$$x = x_0 + \sum_{i=1}^{n} x_i e_i \tag{8.3}$$

where $x_0, x_1, \cdots, x_n \in \mathbb{R}$.

In general, geometric product of the basis elements will generate a new element. The geometric product of two basis vectors e_i and e_j is $e_i e_j$ and it is anti-commutative. Generally, it is denoted

$$e_{ij} = e_i e_j = -e_j e_i = -e_{ji}, \quad \forall i \neq j \tag{8.4}$$

A basis vector multiplies itself to generate two possible values 1 or -1 for non-negative integers p and q such that $n = p + q$. It is formulated

$$e_i{}^2 = \begin{cases} 1, & 1 \leqslant i \leqslant p \\ -1, & p+1 \leqslant i \leqslant n \end{cases} \tag{8.5}$$

For vectors w and z, the geometric product is denoted as:

$$wz = w \cdot z + w \wedge z \tag{8.6}$$

where $w \cdot z$ defines the inner product, $w \wedge z$ denotes the outer product. The geometric product has the properties as follows

$$\begin{cases} w \cdot z = \dfrac{1}{2}(wz + zw) \\ w \wedge z = \dfrac{1}{2}(wz - zw) \end{cases} \tag{8.7}$$

Therefore, the geometric product is an object of mixed grades: it has a scalar part $w \cdot z$ and a bivector part $w \wedge z$.

It can be seen that the geometric product is neither fully symmetric, nor fully anti-symmetric. The mathematical relationship between the two vectors can be obtained completely in GA theory.

In GA, multivectors, which are the extension of vectors to higher dimensions, are the basic elements. Any multivector $\boldsymbol{M} \in (\mathbb{G}_n)^{M \times N}$ is represented as

$$\begin{aligned} \boldsymbol{M} =& E_0(\boldsymbol{M}) + \sum_{1 \leqslant i \leqslant n} E_i(\boldsymbol{M}) e_i + \sum_{1 \leqslant i < j \leqslant n} E_{ij}(\boldsymbol{M}) e_{ij} \\ &+ \cdots + E_{1 \cdots n}(\boldsymbol{M}) e_{1 \cdots n}, \quad E(\boldsymbol{M}) \in \mathbb{R}^{M \times N} \end{aligned} \tag{8.8}$$

In the next section, we will show the basic concept of the reduced geometric algebra to overcome the high complexity issue of the raw geometric algebra.

8.3 Reduced Geometric Algebra (RGA)

The multiplication of GA above is not commutative, which leads to the high algorithm complexity. As an improvement, we provide a novel geometric algebra theory with commutative properties.

8.3.1 Definition

The reduced geometric algebra (RGA) is defined as follows:

$$\gamma_i = \frac{1}{2}(1 + e_i e_{n+i}) \in G_n, \quad i = 1, 2, \cdots, n \tag{8.9}$$

According to Eq.(8.4), the geometric product of γ_i and γ_j is shown as

$$\gamma_i \gamma_j = \gamma_j \gamma_i, \quad i \neq j \tag{8.10}$$

At the same time, we define:

$$\gamma_i{}^2 \triangleq \gamma_i \gamma_j = \begin{cases} \gamma_{i+1}, & i = 1, 2, \cdots, n-1 \\ \gamma_1, & i = n \end{cases} \tag{8.11}$$

Take $n = 3$ for instance, $\gamma_1{}^2{=}\gamma_2, \gamma_2{}^2{=}\gamma_3, \gamma_3{}^2{=}\gamma_1$.

According to Eq.(8.10) and Eq.(8.11), the multiplication of γ_i is commutative. RGA is denoted as $\mathbb{G}_n^R$ and is the space generated by the collection of $\{\{\gamma_1, \gamma_2, \cdots, \gamma_n\}, \{\gamma_{ij} = \gamma_i \gamma_j, 1 \leqslant i \neq j \leqslant n\}\}$. The element k in $\mathbb{G}_2^R$ is denoted as:

$$k = a_1 \gamma_1 + a_2 \gamma_2 + a_{12} \gamma_{12}, \quad a_1, a_2, a_{12} \in \mathbb{R} \tag{8.12}$$

8.3.2 The Properties of Reduced Geometric Algebra

The addition and subtraction operations of $\mathbb{G}_2^R$ can refer to GA, here the multiplication operation is provided as follows. $\forall k, l \in \mathbb{G}_2^R$, suppose $k = a_1\gamma_1 + a_2\gamma_2 + a_3\gamma_{12}$ and $k = a_1\gamma_1 + a_2\gamma_2 + a_3\gamma_{12}$, then the multiplication in $\mathbb{G}_2^R$ are given as:

$$\begin{aligned} k \cdot l &= (a_1\gamma_1 + a_2\gamma_2 + a_3\gamma_{12}) \cdot (b_1\gamma_1 + b_2\gamma_2 + b_3\gamma_{12}) \\ &= (a_1b_3 + a_2b_2 + a_3b_1)\gamma_1 + (a_1b_1 + a_2b_3 \\ &\quad + a_3b_2)\gamma_2 + (a_1b_2 + a_2b_1 + a_3b_3)\gamma_{12} \end{aligned} \tag{8.13}$$

where $\gamma_{12}\gamma_1 = \gamma_1\gamma_2\gamma_1 = \gamma_1\gamma_1\gamma_2 = \gamma_1{}^2\gamma_2 = \gamma_2\gamma_2 = \gamma_1$. similarly, $\gamma_{12}\gamma_2 = \gamma_2$. Moreover, $\gamma_{12}\gamma_{12} = \gamma_1\gamma_2\gamma_1\gamma_2 = \gamma_1{}^2\gamma_2{}^2 = \gamma_1\gamma_2 = \gamma_{12}$.

Form (8.13), we can see that the results of the multiplication of k and l only contain γ_1, γ_2 and γ_{12} components, and no superfluous components are produced. Therefore, the 2-dimensional RGA overcomes the data redundancy embodied in the previous sparse representation for color image based on quaternion.

The norm of the elements in $\mathbb{G}_2^R$ is defined as:

$$\|k\| = \|a\gamma_1 + b\gamma_2 + c\gamma_{12}\| \triangleq |a + b + c| \tag{8.14}$$

The conjugate of k is defined as

$$\bar{k} = a'\gamma_1 + b'\gamma_2 + c'\gamma_{12} \tag{8.15}$$

Then

$$\begin{aligned} k \cdot \bar{k} &= (a\gamma_1 + b\gamma_2 + c\gamma_{12}) \cdot (a'\gamma_1 + b'\gamma_2 + c'\gamma_{12}) \\ &= (ca' + bb' + ac')\,\gamma_1 + (aa' + cb' + bc')\,\gamma_2 + (ba' + ab' + cc')\,\gamma_{12} \\ &= \|k\|^2 \end{aligned} \tag{8.16}$$

According to Eq.(8.16), the following equations are obtained

$$\begin{cases} ca' + bb' + ac' = 0 \\ aa' + cb' + bc' = 0 \\ ba' + ab' + cc' = \|k\|^2 \end{cases} \tag{8.17}$$

Solving the above equations yields the values of the individual components in Eq.(8.15), but not all the elements of $\mathbb{G}_2^R$ are conjugate.

Therefore, the inverse of element k in $\mathbb{G}_2^R$ is denoted as:

$$k^{-1} = \frac{\bar{k}}{\|k\|^2} \tag{8.18}$$

In RGA, multivectors, which are the extension of vectors to higher dimensions, are the basic elements. Any multivector $\boldsymbol{K} \in \left(\mathbb{G}_2^R\right)^{M\times N}$ is so described by

$$\boldsymbol{K} = \boldsymbol{K}_1\gamma_1 + \boldsymbol{K}_2\gamma_2 + \boldsymbol{K}_{12}\gamma_{12} \tag{8.19}$$

where $\boldsymbol{K}_1, \boldsymbol{K}_2, \boldsymbol{K}_{12} \in \mathbb{R}^{M\times N}(0)$.

8.3.3 Singular Value Decomposition (SVD) of RGA

We perform the SVD of RGA using the equivalent complex matrix. Then, the form of (β_1, β_2) is introduced to represent a RGA multivector in another way. (β_1, β_2) contain the following rules:

$$\begin{aligned} \beta_1 &= -\gamma_1 - \gamma_2 + 2\gamma_{12} \\ \beta_2 &= \gamma_1 + \gamma_2 + \gamma_{12} \end{aligned} \tag{8.20}$$

Then, the RGA multivector $\boldsymbol{K}_{(M)}$ is presented as follows:

$$\boldsymbol{K}_{(M)} = \boldsymbol{K}_{1(C)}\beta_1 + \boldsymbol{K}_{2(C)}\beta_2 \tag{8.21}$$

Where

$$\begin{aligned} \boldsymbol{K}_{1(C)} &= \boldsymbol{K}^1_{1(M)}\gamma_1 + \boldsymbol{K}^2_{1(M)}\gamma_2 + \boldsymbol{K}^3_{1(M)}\gamma_{12} \\ &= \boldsymbol{K}^1_{1(M)}\gamma_1 + \left(\boldsymbol{K}^2_{1(M)} + \boldsymbol{K}^3_{1(M)}\gamma_1\right)\gamma_2 \\ &= \boldsymbol{K}^1_{1(C)} + \boldsymbol{K}^2_{1(C)}\gamma_2 \\ \boldsymbol{K}_{2(C)} &= \boldsymbol{K}^1_{2(M)}\gamma_1 + \boldsymbol{K}^2_{2(M)}\gamma_2 + \boldsymbol{K}^3_{2(M)}\gamma_{12} \\ &= \boldsymbol{K}^1_{2(M)}\gamma_1 + \left(\boldsymbol{K}^2_{2(M)} + \boldsymbol{K}^3_{2(M)}\gamma_1\right)\gamma_2 \\ &= \boldsymbol{K}^1_{2(C)} + \boldsymbol{K}^2_{2(C)}\gamma_2 \end{aligned} \tag{8.22}$$

where $\boldsymbol{K}^1_{1(M)}$, $\boldsymbol{K}^2_{1(M)}$, $\boldsymbol{K}^3_{1(M)}$, $\boldsymbol{K}^1_{2(M)}$, $\boldsymbol{K}^2_{2(M)}$ and $\boldsymbol{K}^3_{2(M)}$ are real, γ_1, γ_2 and γ_{12} are basis of RGA. The addition and multiplication of the RGA multivector can be obtained by the corresponding processes of complex matrices.

Inspired by the process of computing the SVD of the quaternion matrix[23,27,35,36], the SVD of the RGA multivector can be performed by using its equivalent complex matrices. Three main steps are included as follows: decomposition, properties of unitary matrices, and reconstruction. We use the form of (β_1, β_2) to simplify these steps.

Suppose a RGA multivector is given as

$$\boldsymbol{K}_{(M)} = A_{(C)} + B_{(C)}\gamma_2 = \boldsymbol{K}_{1(C)}\beta_1 + \boldsymbol{K}_{2(C)}\beta_2 \tag{8.23}$$

Thus, $\boldsymbol{K}_{1(C)}$ and $\boldsymbol{K}_{2(C)}$ are two equivalent complex matrices. The SVD of $\boldsymbol{K}_{1(C)}$ and $\boldsymbol{K}_{2(C)}$ is defined as:

$$\begin{cases} \boldsymbol{K}_{1(C)} = \boldsymbol{U}_{1(C)}\left[\Sigma_{1(C)}\right]\left[\boldsymbol{V}_{1(C)}\right]^H \\ \boldsymbol{K}_{2(C)} = \boldsymbol{U}_{2(C)}\left[\Sigma_{2(C)}\right]\left[\boldsymbol{V}_{2(C)}\right]^H \end{cases} \tag{8.24}$$

where $\Sigma_{1(C)}$ and $\Sigma_{2(C)}$ are two diagonal matrices with real elements, i.e., $\delta_{1,i}$ and $\delta_{2,i}$, $i = 1, \cdots, n$, respectively. The superscript H denotes the Hermitian transpose. The unitary matrices $\boldsymbol{U}_{(C)}$ and $\boldsymbol{V}_{(C)}$ satisfy the following expression:

$$\boldsymbol{U}_{1(C)}\left[\boldsymbol{V}_{1(C)}\right]^H = \boldsymbol{V}_{1(C)}\left[\boldsymbol{V}_{1(C)}\right]^H = \boldsymbol{U}_{2(C)}\left[\boldsymbol{V}_{2(C)}\right]^H = \boldsymbol{V}_{2(C)}\left[\boldsymbol{V}_{2(C)}\right]^H = \boldsymbol{I}_n \tag{8.25}$$

Therefore, RGASVD is defined as

$$\begin{aligned}&\boldsymbol{K}_{(M)}\\&=\left(\boldsymbol{U}_{1(C)}\left[\Sigma_{1(C)}\right]\left[\boldsymbol{V}_{1(C)}\right]^H\right)\beta_1+\left(\boldsymbol{U}_{2(C)}\sum\left[\Lambda_{2(C)}\right]\left[\boldsymbol{V}_{2(C)}\right]^H\right)\beta_2\\&=\left[\boldsymbol{U}_{1(C)}\beta_1+\boldsymbol{U}_{2(C)}\beta_2\right]\left[\Sigma_{1(C)}\beta_1+\Sigma_{2(C)}\beta_2\right]\left\{\left[\boldsymbol{V}_{1(C)}\right]^H\beta_1+\left[\boldsymbol{V}_{2(C)}\right]^H\beta_2\right\}\end{aligned} \tag{8.26}$$

since $\beta_1\beta_2 = 0, {\beta_1}^2 = \beta_1, {\beta_2}^2 = \beta_2$. RGASVD can be simplified as follows:

$$\boldsymbol{K}_{(M)} = \boldsymbol{U}_{(M)}\left[\Sigma_{(M)}\right]\left[\boldsymbol{V}_{(M)}\right]^H \tag{8.27}$$

where $\Sigma_{(M)} = \Sigma_{1(C)}\beta_1 + \Sigma_{2(C)}\beta_2$, $\boldsymbol{U}_{(M)} = \boldsymbol{U}_{1(C)}\beta_1 + \boldsymbol{U}_{2(C)}\beta_2$ and $\boldsymbol{V}_{(M)} = \boldsymbol{V}_{1(C)}\beta_1 + \boldsymbol{V}_{2(C)}\beta_2$.

The SVD of the RGA multivector (RGASVD) is performed by the complex SVD of the equivalent complex matrices. Consequently, the original RGA multivector is reconstructed by the sum of outer products:

$$\boldsymbol{K'}_{(M)} = \boldsymbol{U}_{(M)}\left[\Sigma_{(M)}\right]\left[\boldsymbol{V}_{(M)}\right]^H = \sum_{i=1}^{n} u^i_{(M)}\delta^i_{(M)}\left[v^i_{(M)}\right]^H \tag{8.28}$$

where $\boldsymbol{K'}_{(M)}$ is the reconstruction of $\boldsymbol{K}_{(M)}$. The complexity of RGASVD is much lower than the quaternion application because more real multiplications are needed to calculate the product of two quaternions. Thus, the use of RGASVD for color image processing is more efficient than using quaternion.

8.4 The RGA-sparse Representation Model for Multi-channel Image

Given a color image I, its RGA form is:

$$I(x,y)=I_R(x,y)\gamma_1+I_G(x,y)\gamma_2+I_B(x,y)\gamma_{12} \tag{8.29}$$

where $I_R(x,y)$, $I_G(x,y)$ and $I_B(x,y)$ are the three color components of color image $I(x,y)$ respectively.

Specifically, the three color components are assigned to γ_1, γ_2 and γ_{12} components of the 2-dimensional RGA ($\mathbb{G}_2^R$) respectively, and the color image pixel $I(x,y)$ is represented in the form of a reduced geometric algebraic multivector. Each pixel of the image gives a data structure that is RGA.

To take the inter-relationship among the RGB channels into consideration, a novel sparse representation model (RGA-SR) for color image based on reduced geometric algebra is presented in this chapter. The patch $\boldsymbol{f}$ in the RGA form is given as:

$$\boldsymbol{f} = \boldsymbol{f}_{\mathrm{R}}\gamma_1 + \boldsymbol{f}_{\mathrm{G}}\gamma_2 + \boldsymbol{f}_{\mathrm{B}}\gamma_{12}, \quad \boldsymbol{f} \in \left(\mathbb{G}_2^R\right)^N \tag{8.30}$$

Then, the proposed model is defined as:

$$\min_{\boldsymbol{g}} \|\boldsymbol{g}\|_0, \quad \text{s.t.} \quad \boldsymbol{f} = \boldsymbol{D}\boldsymbol{g} \tag{8.31}$$

where $\boldsymbol{D} = \boldsymbol{D}_{\mathrm{R}}\gamma_1 + \boldsymbol{D}_{\mathrm{G}}\gamma_2 + \boldsymbol{D}_{\mathrm{B}}\gamma_{12} \in \left(\mathbb{G}_2^R\right)^{N\times M}$ is the RGA dictionary consisting M RGA atoms and $\boldsymbol{g} = \boldsymbol{g}_1\gamma_1 + \boldsymbol{g}_2\gamma_2 + \boldsymbol{g}_3\gamma_{12} \in \left(\mathbb{G}_2^R\right)^M$ is denoted as a RGA coefficient vector. The objective function $\|\boldsymbol{g}\|_0$ obtains the amount of non-zero elements of $\boldsymbol{g}$.

Then, we can obtain the following equation as the generalized form of color image representation model

$$[\boldsymbol{f}_{\mathrm{R}}, \ \boldsymbol{f}_{\mathrm{G}}, \ \boldsymbol{f}_{\mathrm{B}}] = [\boldsymbol{D}_{\mathrm{R}}, \ \boldsymbol{D}_{\mathrm{G}}, \ \boldsymbol{D}_{\mathrm{B}}]\tilde{\boldsymbol{g}} \tag{8.32}$$

Since

$$\begin{aligned}\boldsymbol{f} =& \boldsymbol{D}\boldsymbol{g} \Leftrightarrow \boldsymbol{f}_{\mathrm{R}}\gamma_1 + \boldsymbol{f}_{\mathrm{G}}\gamma_2 + \boldsymbol{f}_{\mathrm{B}}\gamma_{12} \\ &= (\boldsymbol{D}_{\mathrm{R}}\gamma_1 + \boldsymbol{D}_{\mathrm{G}}\gamma_2 + \boldsymbol{D}_{\mathrm{B}}\gamma_{12})(\boldsymbol{g}_1\gamma_1 + \boldsymbol{g}_2\gamma_2 + \boldsymbol{g}_3\gamma_{12}) \\ &= (\boldsymbol{D}_{\mathrm{R}}\boldsymbol{g}_3 + \boldsymbol{D}_{\mathrm{G}}\boldsymbol{g}_2 + \boldsymbol{D}_{\mathrm{B}}\boldsymbol{g}_1)\gamma_1 + (\boldsymbol{D}_{\mathrm{R}}\boldsymbol{g}_1 + \boldsymbol{D}_{\mathrm{G}}\boldsymbol{g}_3 + \boldsymbol{D}_{\mathrm{B}}\boldsymbol{g}_2)\gamma_2 \\ &\quad + (\boldsymbol{D}_{\mathrm{R}}\boldsymbol{g}_2 + \boldsymbol{D}_{\mathrm{G}}\boldsymbol{g}_1 + \boldsymbol{D}_{\mathrm{B}}\boldsymbol{g}_3)\gamma_{12}\end{aligned} \tag{8.33}$$

Therefore, we can obtain

$$\begin{cases} \boldsymbol{f}_{\mathrm{R}} = \boldsymbol{D}_{\mathrm{R}}\boldsymbol{g}_3 + \boldsymbol{D}_{\mathrm{G}}\boldsymbol{g}_2 + \boldsymbol{D}_{\mathrm{B}}\boldsymbol{g}_1 \\ \boldsymbol{f}_{\mathrm{G}} = \boldsymbol{D}_{\mathrm{R}}\boldsymbol{g}_1 + \boldsymbol{D}_{\mathrm{G}}\boldsymbol{g}_3 + \boldsymbol{D}_{\mathrm{B}}\boldsymbol{g}_2 \\ \boldsymbol{f}_{\mathrm{B}} = \boldsymbol{D}_{\mathrm{R}}\boldsymbol{g}_2 + \boldsymbol{D}_{\mathrm{G}}\boldsymbol{g}_1 + \boldsymbol{D}_{\mathrm{B}}\boldsymbol{g}_3 \end{cases} \tag{8.34}$$

Then, we rewrite Eq.(8.34) in the same form as Eq.(8.32) to obtain

$$\tilde{g} = [g_{\mathrm{R}} \quad g_{\mathrm{G}} \quad g_{\mathrm{B}}] = \begin{bmatrix} g_3 & g_2 & g_1 \\ g_1 & g_3 & g_2 \\ g_2 & g_1 & g_3 \end{bmatrix} \tag{8.35}$$

The advantages of reduced geometric algebra (RGA)-based sparse representation model in Eq.(8.31) over traditional sparse models for color image can be summarized as follows.

First, both the orthogonal property and the correlation among multiple channels are jointly preserved in the coefficient matrix. Each color channel is linearly correlated with the RGA dictionary, which is superior to traditional models in which atoms are selected from three independent channel dictionaries since the three-channel interrelationship for color patches can be preserved by properly training the RGA-based dictionary $\boldsymbol{D}$.

Then, as shown in Eq.(8.34), the explicit linear relationship among channel dictionaries, which has been proven to be useful in color constancy, is described by $\boldsymbol{g}_1$, $\boldsymbol{g}_2$ and $\boldsymbol{g}_3$.

Finally, since the RGA is commutative, the complexity of the algorithm is effectively reduced during the dictionary training stage, and the data redundancy is reduced accordingly.

8.5 RGA-based Dictionary Training

8.5.1 RGA Dictionary Training

When both the dictionary and the coefficients are unknown variables, the RGA-based dictionary training process can be deemed as an extension of the model in Eq.(8.31). Then, this process can be defined as

$$\{\hat{\boldsymbol{D}} \quad \hat{\boldsymbol{G}}\} = \arg\min_{\boldsymbol{D},\boldsymbol{G}} \|\boldsymbol{F} - \boldsymbol{D}\boldsymbol{G}\|_F^2 + \gamma\|\boldsymbol{G}\|_0 \tag{8.36}$$

where $\boldsymbol{F}=\{\boldsymbol{f}_i, 1 \leqslant i \leqslant K\} \in \left(\mathbb{G}_2^R\right)^{N\times K}$ is the set of color image patches, $\boldsymbol{D} = \{\boldsymbol{d}_i, 1 \leqslant i \leqslant M\} \in \left(\mathbb{G}_2^R\right)^{N\times M}$ denotes the dictionary with M atoms in the RGA form, and $\boldsymbol{G}=\{\boldsymbol{g}_i, 1 \leqslant i \leqslant K\} \in \left(\mathbb{G}_2^R\right)^{M\times K}$ is the sparse coefficient vector, and $\|\boldsymbol{G}\|_0$ counts the number of nonzero elements for each column in $\boldsymbol{G}$. As the extension of K-SVD[2], a dictionary training method based on RGA theory, namely K-RGASVD, is introduced for the first time to realize the optimized RGA-based dictionary. It is composed of the following two parts, sparse coding phase and dictionary updating phase.

First, a sparse coefficient matrix $\boldsymbol{G}$ needs to be obtained during the sparse coding stage, in which a fixed dictionary $\boldsymbol{D}$ in Eq.(8.36) is already given. In general, there exist a variety of methods that can be used to solve this sparse coding problem. Among them, methods such as matching pursuit (MP)[37], basis pursuit (BP)[38] and orthogonal matching pursuit (OMP)[23,27] are the most common ones. Inspired by the high efficiency of OMP algorithm, the reduced geometric algebra-based orthogonal matching pursuit algorithm called RGA-OMP is designed in this chapter.

The RGA-OMP algorithm can be applied to solve the problem of decomposing signal $\boldsymbol{F} \in \left(\mathbb{G}_2^R\right)^{N\times K}$ with a fixed RGA-based dictionary $\boldsymbol{D} \in \left(\mathbb{G}_2^R\right)^{N\times M}$, such that

$$\boldsymbol{G} = \arg\min_{\boldsymbol{G}} \|\boldsymbol{F} - \boldsymbol{D}\boldsymbol{G}\|_2^2, \quad \text{s.t.} \quad \|\boldsymbol{G}\| \leqslant T \tag{8.37}$$

Where $\boldsymbol{G} \in \left(\mathbb{G}_2^R\right)^{M\times K}$ is the sparse coefficient vector and $\|\boldsymbol{G}\| \leqslant T$ is the stopping criteria. By specifying the maximum number of non-zero coefficients per signal, it can mitigate the NP-hard l_0-norm sparse coding problem.

Table 8.1 shows the details of RGA-OMP algorithm for each color image patch.

Table 8.1 The RGA-based orthogonal matching pursuit algorithm

Input: $\boldsymbol{f} \in \left(\mathbb{G}_2^R\right)^N$, $\boldsymbol{D} \in \left(\mathbb{G}_2^R\right)^{N\times M}$ Output: $\boldsymbol{g} \in \left(\mathbb{G}_2^R\right)^M$
1. Initialization: $i = 1$, $\boldsymbol{e}^0 = \boldsymbol{f}$, active dictionary; $\boldsymbol{D}^0 = \varnothing$ 2. Repeat for $\boldsymbol{g} \in \left(\mathbb{G}_2^R\right)^M$ do $c_m^i \leftarrow \boldsymbol{d}\left(\mathbb{G}_2^R\right)_m \boldsymbol{e}^{i-1}$ end for selection: $m^i \leftarrow \arg\max_m \left\|c_m^i\right\|$ Active dictionary: $\boldsymbol{D}^i \leftarrow \left[\boldsymbol{D}^{i-1}, \boldsymbol{d}_{m^i}\right]$ Active coefficients: $\boldsymbol{g}^i \leftarrow \arg\max_{\boldsymbol{g}} \left\|\boldsymbol{f} - \boldsymbol{D}^i\boldsymbol{g}\right\|_2^2$ Residue: $\boldsymbol{e}^i \leftarrow \boldsymbol{f} - \boldsymbol{D}^i\boldsymbol{f}^i$, $i \leftarrow i+1$ until stopping criterion

Obviously, RGA-OMP for RGA multivector is an extension of traditional OMP for the real data. Generally, the increase of the iteration number J leads to the consequent reduction of the reconstruction residual. And since the sparse coding phase aims to obtain the value of reconstruction residual as small as possible, and the sparsity of coefficient vector as large as possible, we set an upper bound of J to balance them.

Once the solution of sparse coding problem is given, the next step is to train the RGA-based dictionary $\boldsymbol{D}$. Instead of fixing the coefficient vector during dictionary learning, our K-RGASVD method has the ability to update coefficient simultaneously, which is highly efficient. For each atom $\boldsymbol{d}_m$ and the corresponding coefficients $\boldsymbol{G}^m$-the m-th row of $\boldsymbol{G}$, we update both of them by decomposing the remaining representation error $\boldsymbol{E}_m = \boldsymbol{F} - \sum\limits_{j \neq m} \boldsymbol{d}_j \boldsymbol{g}^j$ using RGASVD presented in section 8.3. As mentioned above, it can be observed that there are more information contained in the first basis of RGASVD than in the SVD of color images, which shows the more efficiency of using K-RGASVD to update atoms and corresponding coefficients. The details of K-RGASVD algorithm is given in Table 8.2.

Table 8.2 The RGA-based dictionary learning using K-RGASVD method

Parameters: K-block size, J-number of training iterations, M-dictionary size
1. Initialization: Training data: $\boldsymbol{F}=\{\boldsymbol{f}_i, 1 \leqslant i \leqslant K\} \in (\mathbb{G}_2^R)^{N \times K}$, the RGA dictionary: $\boldsymbol{D}=\{\boldsymbol{d}_i, 1 \leqslant i \leqslant M\} \in (\mathbb{G}_2^R)^{N \times M}$, and is random samples from $\boldsymbol{F}$.
2. For $j = 1, \cdots, J$ do:
(1) Sparse coding phase: Compute the sparse coefficient matrix $\boldsymbol{G}=\{\boldsymbol{g}_i, 1 \leqslant i \leqslant K\} \in (\mathbb{G}_2^R)^{M \times K}$ via RGA-OMP.
(2) Dictionary update phase: For each RGA dictionary atom $\boldsymbol{d}_m$ in $\boldsymbol{D}^J$ do.
(i) Index: $\mu_m = \{i \,\vert 1 \leqslant i \leqslant N, \boldsymbol{G}(m,i) \neq 0\}$, where $\boldsymbol{G}(m,i)$ indicates the carry at i-th row and m-th column of coefficient matrix $\boldsymbol{G}$.
(ii) Error: $\boldsymbol{E}_m \leftarrow \boldsymbol{F} - \sum\limits_{j \neq m} \boldsymbol{d}_j \boldsymbol{g}^j$
(iii) Select the columns corresponding to μ_m using RGASVD. $$\boldsymbol{E}_m^R = \boldsymbol{E}_m(i,j)\vert_{j=\mu_m} = \boldsymbol{U} \wedge \boldsymbol{V}^H$$
(iv) $\boldsymbol{d}_m \leftarrow \boldsymbol{U}(:,1)$ and $\boldsymbol{g}_R^m = \mu_m^{\mathrm{T}} \boldsymbol{g}^m \leftarrow \boldsymbol{V}(1,1)^H \cdot \wedge(1,1)$.

8.5.2 Further Analysis

To further analyze the efficiency of the RGA-based joint sparse representation model, the time complexity of K-RGASVD will be compared to K-SVD[2] (the traditional monochrome sparse model) and K-QSVD[23] (quaternion-based sparse model). Because K-RGASVD makes use of the same framework of traditional K-SVD and quaternion form K-QSVD, its convergence is also similar to them. In each iteration, these three methods all consist of sparse coding phase and dictionary updating phase. In sparse coding phase, we use RGA-OMP to obtain sparse codes while keeping the dictionary fixed. The RGA-OMP is also a greedy algorithm like OMP, which means the construction residual will decrease when the number of nonzero coefficients we select keep growing. That is to say, the reconstruction residual reduces regarding the number of iteration. In dictionary learning phase, we optimize dictionary as K-SVD does while keeping the sparse codes fixed. The energy of the previous residual error matrix is reduced while each atom updating.

Dictionaries learned with K-SVD, K-QSVD and K-RGASVD are visualized in Fig.8.1.

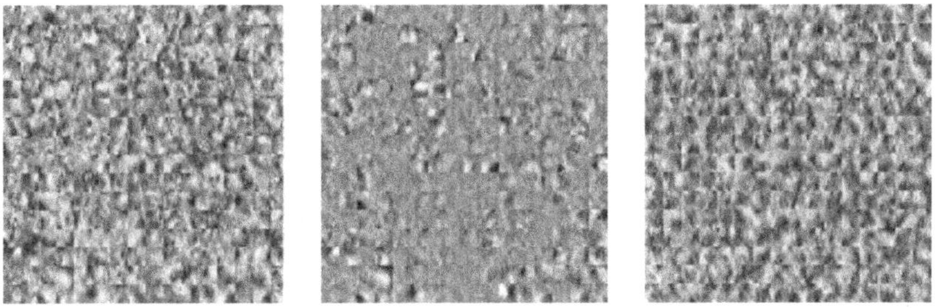

(a) K-SVD (b) K-QSVD (c) K-RGASVD

Fig. 8.1 Visualization of the color image dictionaries learned by K-SVD (a), K-QSVD (b), and K-RGASVD (c)

It is clearly shown in Fig.8.1 that the dictionary learned by the K-SVD has many color-less atoms for the reconstruction of spatial structures. Because of the emphasis on the basic spatial structures of concatenated channel images by K-SVD method, it cannot generate atoms rich enough to represent the diversity of colors. In contrast, the quaternion dictionary earned by the K-QSVD has more color which captures the interrelationship among color channels as well as the spatial coherence better. However, its main drawback is exposed while processing the real part of the quaternionic data. For the quaternion-based sparse model, although the image $\dot{\boldsymbol{f}} \in H^N$ and the atoms $\dot{\boldsymbol{D}}$ are pure, the obtained sparse approximation $\dot{\boldsymbol{f}} = \dot{\boldsymbol{D}}\dot{\boldsymbol{a}}$ is not necessarily pure. Its real part has to be reset, which increases the residue and then leads to the data redundancy.

In order to advance the above two dictionaries, the RGA dictionary learned by the K-RGASVD method is obtained and shown in Fig.8.1(c). The color of the RGA dictionary seems to be nearly equivalent, compared to the quaternion dictionary learned by the K-QSVD method in Fig.8.1 (b). The proposed RGA- based sparse model takes full advantage of the RGA (only three base elements $\gamma_1, \gamma_2, \gamma_{12}$ are used for color image), no extra component is produced. Thus, the RGA overcomes the data redundancy experienced by the previous quaternion in color image modeling.

8.6 Experiments and Analysis

In this section, we conduct color image reconstruction and denoising experiments to evaluate the practical performance and computational efficiency of our proposed RGA-based sparse representation model.

8.6.1 Data Sets

To better evaluate the adaptive capacity, variety of color images are used and included in the color image data set, some of which can be obtained from http://tabby.vision.mcgill.ca/html/browsedownload.html. Fig.8.2 shows the example color images used in the experiments.

Fig. 8.2 Examples of training images for color image denoising and reconstruction

8.6.2 Color Image Reconstruction

We first evaluate the proposed K-RGASVD based sparse model through color image reconstruction experiments at different sparsity, comparing with K-SVD and quaternion-based sparse models. Based on the data sets above, the dictionaries using K-SVD, K-QSVD and K-RGASVD are trained respectively on the same training samples.

A reasonable computational complexity can be maintained by choosing an appropriate number of dictionary atoms, which is not particularly large or small. In the sparse coding stage, we set the sparsity parameter T in OMP/QOMP/RGA-OMP for the three sparse models, where T means the maximum value of nonzero coefficients allowed for representing each image block.

The first reconstruction experiment is implemented to verify the superior performance of our K-RGASVD based sparse model through the PSNR (dB) values over different sparse parameter T for the three sparse models. And the number of the atoms of the three dictionary have been all fixed to be $M = 256$. The relationship between reconstructed PSNR and sparse sparsity parameter T is shown in Fig.8.3.

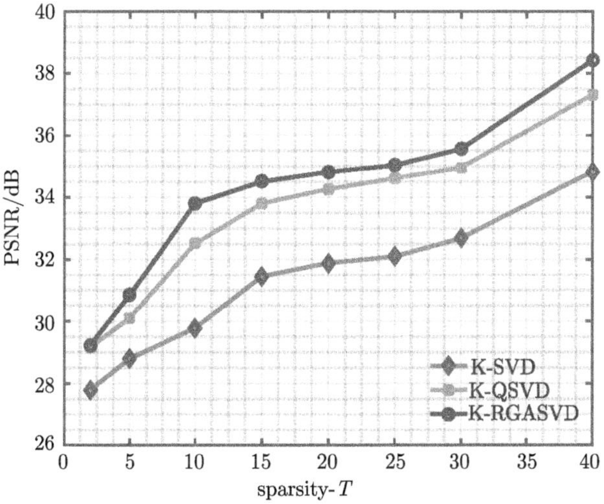

Fig. 8.3 Comparison of K-RGASVD, K-QSVD and K-SVD based sparse model for color image reconstruction-PSNR values vs. sparse parameter T

It can be seen from Fig.8.3, the PSNR values of the proposed RGA-based and quaternion-based sparse models are almost similar or a little higher. And they can present exactly higher PSNR values comparing to the traditional sparse model using K-SVD method under the same sparse parameter. When the number of atoms used increases, the superiority becomes even more distinct.

Besides, the number of dictionary atoms is also an important indicator to evaluate the performance of reconstruction, and normally, has been fixed to be $M = 256$. Since we suspect quaternions and our RGA better capture the correlations of color channels, we now test the K-SVD, K-QSVD and our K-RGASVD algorithms with variety of atoms to see if these correlations can lead to a lower-redundancy for the dictionary. The results are depicted in Fig. 8.4.

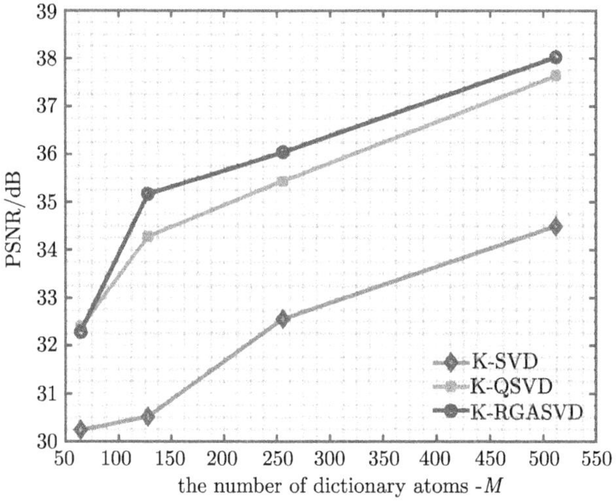

Fig. 8.4 Comparison of K-RGASVD, K-QSVD and K-SVD sparse model for color image reconstruction-PSNR values vs. the number of dictionary atoms M

In this experiment, we test the reconstruction results of the three methods using dictionaries with the same number of atoms. The number of the atoms is set as 64, 128, 256, and 512. Clearly, in Fig.8.4, compared with the K-SVD based model, the K-QSVD and our K-RGASVD based models show great improvement in terms of the reconstruction PSNR values. And the increasing number of dictionary atoms greatly improves the results. This improvement can be more effective in the adaptive dictionary method.

Generally, our K-RGASVD based model shows similar or a little higher PSNR values compared with the K-QSVD based model.

8.6.3 Color Image Denosing

Then, another application-denoising experiments are conducted on various color images at different noise levels. As described above, each pixel of a color image can be represented in the RGA form using the proposed algorithm. Let $\boldsymbol{F}_0$ be a clean color image in RGA form, consider a noisy version of it, $\boldsymbol{Y}$, corrupted by an additive zero-mean white Gaussian noise in RGA form with standard deviation σ. Then the noisy image can be represented as

$$\boldsymbol{Y} = \boldsymbol{F}_0 + \boldsymbol{v} \tag{8.38}$$

where $\boldsymbol{F}_0=\{\boldsymbol{f}_{0i} \in \left(\mathbb{G}_2^R\right)^N, 1 \leqslant \mathrm{i} \leqslant K\} \in \left(\mathbb{G}_2^R\right)^{N\times K}$ and $\boldsymbol{f}_{0i}$ are overlapped patches of the clean color image $\boldsymbol{F}_0$. We consider overlapped image patches of size $\sqrt{N} \times \sqrt{N}$ pixels, ordered lexicographically as column vectors $\boldsymbol{y} \in \left(\mathbb{G}_2^R\right)^N$, and $\boldsymbol{Y}=\{\boldsymbol{y}_i \in \left(\mathbb{G}_2^R\right)^N, 1 \leqslant i \leqslant K\} \in \left(\mathbb{G}_2^R\right)^{N\times K}$.

The denoising problem can be transformed to the solution of the following function:

$$\left\{\hat{\boldsymbol{D}}, \hat{\boldsymbol{g}}_{kl}, \hat{\boldsymbol{F}}\right\} = \min_{\hat{\boldsymbol{D}}, \hat{\boldsymbol{g}}_{kl}, \hat{\boldsymbol{F}}_0} \gamma \|\boldsymbol{Y} - \boldsymbol{F}\|_2^2 + \sum_{k,l} \eta_{kl} \|\boldsymbol{g}_{kl}\|_0 + \sum_{k,l} \|\boldsymbol{D}\boldsymbol{g}_{kl} - J_{kl}\boldsymbol{Y}\|_2^2 \tag{8.39}$$

where $\hat{\boldsymbol{D}}$, $\hat{\boldsymbol{g}}_{kl}$ and $\hat{\boldsymbol{F}}$ are the estimation of dictionary $\boldsymbol{D} \in \left(\mathbb{G}_2^R\right)^{N\times M}$, sparse coefficient vector $\boldsymbol{g}_{kl}$ and color image $\boldsymbol{F}_0$, respectively. The indices $[k,l]$ denote the location of the patch, J_{kl} is the operator extracting the image patch at coordinates $[k,l]$ from $\boldsymbol{F}_0$, and the vector $\boldsymbol{g}_{kl} \in \left(\mathbb{G}_2^R\right)^{M\times 1}$ is the coefficient vector for the patch at coordinates $[k,l]$. The first term in Eq.(8.39) imposes the likelihood that demands proximity between $\boldsymbol{Y}$ and $\boldsymbol{F}$. The latter two terms enforce the image prior, assuming each patch admit sparse representation without noise over dictionary $\boldsymbol{D}$.

The solution of Eq.(8.39) can be obtained by the proposed model, where the RGA-OMP is implemented to suppressed noise points of color image.

Generally, we set, the number of iterations $J = 30$, the size of color image patches to 8×8, the threshold of denoising $\varepsilon = N(C\sigma)^2$ and $C = 1.15$, the number of RGA dictionary atoms $M = 256$.

Fig.8.5 shows the visualize comparison of the three sparse methods for color image denoising. The color images chosen from Fig.8.2 are contaminated by additive Gaussian white noise with standard deviation 20 ($\sigma = 20$). It is clearly shown that the K-SVD based model loses some color information and structures owing to the loss of correlation of RGB channels. Also, it tends to introduce color bias and blurring effects. As the improvement, the K-QSVD based model treats the total color image as a quaternion matrix and analogously, the K-RGASVD based model regards the image to be a multivector in the RGA form. In terms of denoising effect, the two models seem to be slightly equivalent. Particularly, the advantages are shown in homogenous region, not only more of the image particulars are presented, but also colors with better fidelity can be preserved after denoising.

(a) Original images (b) Noisy images (σ=20) (c) K-SVD (d) K-QSVD (e) K-RGASVD

Fig. 8.5 Denoising results

The PSNR values using different denoising models for different color testing images over different noise levels arc set out in Table 8.3.

Table 8.3 PSNR values (dB) obtained by different denoising methods at each noisy level

σ	methods	beach	fall	butterfly	flower	parrot	bird
10	K-SVD	32.05	31.24	31.27	32.73	32.70	32.32
	K-QSVD	34.46	36.65	32.14	34.11	37.51	33.28
	K-RGASVD	**34.59**	**37.42**	**35.08**	**34.31**	**38.13**	**36.31**
20	K-SVD	29.17	30.47	29.72	30.45	30.40	30.20
	K-QSVD	32.62	31.66	30.41	31.48	31.66	31.37
	K-RGASVD	**33.82**	**33.16**	**31.47**	**31.45**	**32.16**	**31.92**
30	K-SVD	30.83	32.81	28.41	30.00	30.31	27.90
	K-QSVD	**33.82**	34.02	30.17	31.73	30.27	33.28
	K-RGASVD	32.35	**34.71**	**31.54**	**33.61**	**32.11**	**33.82**
40	K-SVD	28.36	26.89	23.08	27.10	26.04	27.88
	K-QSVD	**29.75**	27.87	25.98	27.66	27.71	28.63
	K-RGASVD	29.28	**27.26**	**26.34**	**28.16**	**28.17**	**29.40**

The noisy images are obtained by adding white Gaussian noise with standard deviation $\sigma = 10, 20, 30, 40$. These quantitative evaluation results further illustrate the conclusion in an objective way. It is obviously observed that, from Table 8.3, the use of RGA indeed captures the relationship of the color channels, resulting in quite better results than the first scheme that uses K-SVD method for each channel separately. In terms of PSNR values, the PSNR values of the image using K-RGASVD method give almost as much high performance as with the K-QSVD method. Also, the PSNR values of these two models resemble the one based on K-SVD method.

To further analyze the efficiency of our proposed K-RGASVD based model, we study the computational complexity of this algorithm and compare it to K-QSVD based model. The approximate time is calculated by using MATLAB on a desktop PC with Intel ® Core™ i3-2328M 2.20GHz CPU and 4GB memory. Table 8.4 shows that the elapsed time of K-RGASVD is less than the elapsed time of K-QSVD.

Compared with K-QSVD algorithm in quaternion-based sparse model, our K-RGASVD algorithm has much lower computational complexity. Moreover, focusing on color image representation, in quaternion-based sparse model, the quaternion addition and multiplication require 4 FLOPs and 28 FLOPs (16 floating-point multiplications and 12 floating-point additions), respectively[23]. And in our RGA-based sparse model, the RGA addition and multiplication only require 3 FLOPs and 15 FLOPs (9 floating-point multiplications and 6 floating-point additions). Furthermore, the multiplication operation of the RGA satisfies the commutation rule, so that the computational complexity is much lower than quaternion in the concrete programming process.

Table 8.4 Computation time (s) for different denoising methods with $\sigma = 20$

methods	beach	fall	butterfly
K-QSVD	1528.42	1708.10	1638.52
K-RGASVD	438.36	481.14	468.16
methods	flower	parrot	bird
K-QSVD	1727.88	1548.10	1634.56
K-RGASVD	482.07	439.75	463.43

The computational complexity of the proposed algorithm can be ascribed to the parameter estimation of the testing image, the eigen decomposition of the RGA, and the multiplication of two RGA unitary matrix operations.

From Table 8.4, we can clearly see that, owing to the commutative multiplication operation of RGA, our proposed K-RGASVD algorithm achieves far less computational complexity compared to the K-QSVD algorithm.

Therefore, we can conclude that, compared with the existing sparse models, our proposed joint representation model based on RGA can greatly improve the performance of color image processing while reducing the time complexity and removing the data redundancy.

8.7 Conclusions

In this chapter, a novel sparse representation model for multi-channel image based on reduced geometric algebra is proposed, which formulates the color image as a multivector with the spatial and spectral information in RGA space. Our proposed K-RGASVD based model is shown better at retaining certain texture and color information in the image for the retention of the correlation between multiple RGB channels. The experiments of reconstruction and denoising on natural color images demonstrate our proposed model can preserve inherent color structures completely, remove the data redundancy and reduce the computational complexity.

Currently, our model focuses on modeling the multi-channel correlation which can be regarded as a sequence of channels for image information. We would like to extend to the case covering multiple images in video. On the other hand, we want to investigate the multivector sparse representation of multispectral images based on geometric algebra. It is expected that the proposed RGA-SR model will be a homogeneous and efficient tool in various applications of image analysis.

References

[1] Elad M, Aharon M. Image denoising via sparse and redundant representations over learned dictionaries. IEEE Trans. Image Process, 2006, 15(12): 68-80.

[2] Aharon M, Elad M, Bruckstein A. The K-SVD: An algorithm for designing of overcomplete dictionaries for sparse representations. IEEE Trans. Image Process, 2006, 54(11): 4311-4322.

[3] Carmeli A, Turek J. Quaternion K-SVD for color image denoising. Geometric Image Process. Technion-Israel Institute of Technology, Tech. Rep., 2013.

[4] Yu M Q, Xu Y, Sun P. Single color image super-resolution using quaternion-based sparse representation. in Proc. IEEE Int. Conf. Acoust., Speech Signal Process. (ICASSP), 2014: 5804-5808.

[5] Chan W L, Choi H, Baraniuk R G. Coherent multiscale image processing using dual-tree quaternion wavelets. IEEE Trans. Image Process, 2008, 17(7): 1069-1082.

[6] Do M N, Vetterli M. Pyramid directional filter banks and curvelets. Proc. IEEE Int. Conf. on Image Processing (ICIP), 2001: 158-161.

[7] Do M N, Vetterli M. Framing pyramids. IEEE Trans. Signal Process, 2003, 51(9): 2329-2342.

[8] Guo L Q, Zhu M. Quaternion Fourier–Mellin moments for color images. Int. J. Pattern Recognit., 2011, 44(2): 187–195.

[9] Burt P J, Adelson E H. The Laplacian pyramid as a compact image code. IEEE Trans. Commun., 1983, 31(4): 532-540.

[10] Wang Z C, Jiang R J, Jiang X F, et al. Learning sparse representations by K-SVD for facial expression classification. 2015 4th International Conference on Computer Science and Network Technology (ICCSNT), 2015: 772-775.

[11] Wright J, Ma Y, Mairal J, et al. Sparse representation for computer vision and pattern recognition. Proc. IEEE, 2010, 98(6): 1031-1044.

[12] Jia L, Song S T, Yao L H, et al. Image denoising via sparse representation over grouped dictionaries with adaptive atom size. IEEE Access, 2017, 5: 22514-22529.

[13] Chen F, Yu H M, Hu R. Shape sparse representation for joint object classification and segmentation. IEEE Trans. Image Process., 2013, 22(3): 992-1004.

[14] Agarwal S, Roth D. Learning a sparse representation for object detection.European Conf. on Computer Vision (ECCV), 2006: 97-101.

[15] Mairal J, Elad M, Sapiro G. Sparse representation for color image restoration.IEEE Trans. Image Process, 2007, 17(1): 53-69.

[16] Pang J, Au O C, Tang K, et al. Image colorization using sparse representation. in Proc. IEEE Int. Conf. Acoust., Speech, Signal Process (ICASSP), 2013: 1578-1582.

[17] Li X, Gunturk B, Zhang L. Image demosaicing: A systematic survey. Proc. SPIE 2008, 6822: 68221J.

[18] Hamilton W R. Elements of quaternions. Int. J. Sci., 1901, 14(75): 65-66.

[19] Kolaman A, Yadid-Pecht O. Quaternion structural similarity: A new quality index for color images. IEEE Trans. Image Process, 2011, 21(4): 1526-1536.

[20] Augereau B, Carré P. Hypercomplex polynomial wavelet packet application for color image. in Proc. Conf. Appl. Geometric Algebras Comput. Sci. Eng. (AGACSE), 2012.

[21] Choukroun D, Weiss H, Bar-Itzhack I Y, et al. Quaternion estimation from vector observations using a matrix Kalman filter. IEEE Trans. Aerosp. Electron. Syst., 2012, 48(4): 3133-3158.

[22] Yu L C, Xu Y, Xu H T, et al. Quaternion-based sparse representation of color image. in Proc. IEEE Int. Conf. Multimedia Expo (ICME), 2013: 1-7.

[23] Xu Y, Yu L C, Xu H T, et al. Vector sparse representation of color image using quaternion matrix analysis. IEEE Trans. Image Process, 2015, 24(4): 1315-1328.

[24] Gai S, Yang G, Wan M, et al. Denoising color images by reduced quaternion matrix singular value decomposition. Multidimensional Systems and Signal Processing, 2015.

[25] Gai S, Wan M, Wang L, et al. Reduced quaternion matrix for color texture classification. Neural Computing & Applications, 2014, 25(3-4): 945-954.

[26] Gai S, Yang G, Wan M, et al. Sparse representation based on vector extension of reduced quaternion matrix for multiscale image denoising. IET Image Processing, 2016, 10(8): 598-607.

[27] Barthélemy Q, Larue A, Mars J I. Color sparse representations for image processing:

Review, models, and prospects. IEEE Trans. Image Process, 2015, 24(11): 3978-3989.

[28] Arsenovic A. Applications of conformal geometric algebra to transmission line theory. IEEE Access, 2017, 5: 19920-19941.

[29] Franchini S, Gentile A, Sorbello F, et al. Conformal ALU: A conformal geometric algebra coprocessor for medical image processing. IEEE Transactions on Computers, 2015, 64(4): 955-970.

[30] Batard T, Berthier M, Saint-Jean C. Clifford fourier transform for color image processing. In: Bayro-Corrochano E, Scheuermann G. (eds.) Geometric Algebra Computing, Engineering and Computer Science, 2010: 135-161.

[31] Roy S, Mitra A, Setua S K. Color & Grayscale Image Representation Using Multivector. Proceedings of the 2015 Third International Conference on Computer, Communication, Control and Information Technology (C3IT), 2015: 1-6.

[32] Oviedo-Barriga J, Carbajal-Espinosa O, González-Jiménez L, et al. Robust Tracking of Bio-inspired References for a Biped Robot Using Geometric Algebra and Sliding Modes. 2013 IEEE International Conference on Robotics and Automation, 2013: 5317-5322.

[33] Zou L, Lasenby J, He Z H. Beamforming with distortionless co-polarisation for conformal arrays based on geometric algebra. IET Radar, Sonar & Navigation, 2011, 5(8): 842-853.

[34] Li Y S, Liu W M, Li X T, et al. GA-SIFT: A new scale invariant feature transform for multispectral image using geometric algebra. Information Sciences, 2014, 281: 559-572.

[35] Bihan N L, Sangwine S J. Jacobi method for quaternion matrix singular value decomposition. Appl. Math. Comput., 2007, 187(2): 1265–1271.

[36] Kolaman A, Yadid-Pecht O. Quaternion structural similarity: A new quality index for color images. IEEE Trans Image Process, 2011, 21(4): 1526–1536.

[37] Shahar S S, Tabrikian J. Efficient computation of MSE lower bounds via matching pursuit. IET Radar, Sonar & Navigation, 2017, 24(12): 1798-1802.

[38] Cattenoz M, Savy L, Marcos S. Extended orthogonal matching pursuit for robust and fast target localisation in multiple-input multiple-output radar. IET Radar, Sonar & Navigation, 2017, 11(11): 1709-1717.

Chapter 9

An Extended Multilayer Perceptron Model Using Reduced Geometric Algebra

9.1 Introduction

Multilayer perceptron, known as multi layer feed-forward neural network, is the most well-known artificial neural network (ANN) model. On account of its good nonlinear mapping ability, it has been widely applied in variety of applications like pattern recognition, image processing, fusion approximation, optimization computation, and so on[1−4]. MLP models achieve mappings from input space to output space, and implement its training utilizing mainly back propagation (BP) algorithm to accommodate the connection weights of the network, and are commonly applied in supervised learning tasks[5,6]. Recently, considerable researches has emerged towards both the optimization and extended applications of MLP models. In Ref.[7] Rossi et al. proposed an extension of the multilayer perceptron (MLP) that works on functional inputs, and then they subsequently proposed to rely on a representation of input and weight functions by projection on a truncated base[8]. Recently, Zhang et al.[9] extended MLP to a new MLP-like neural network and has demonstrated to require fewer neural nodes and adjustable parameters in the same situation. Then, Yang et al.[10] analyzed the sensitivity of a split-complex multilayer perceptron (split-CMLP) due to the errors of the inputs and the connection weights between neurons, and showed that the sensitivity is affected by the number of the layers and the number of the neurons adopted in each layer. Gas[5] gave an extension of a self-organizing map called self-organizing multilayer perceptron whose purpose was to achieve quantization of spaces of functions. Grippo et al.[11] considered the learning problem of multilayer perceptrons (MLPs) formulated as the problem of minimizing a smooth error function.

Nevertheless, there does not exist efficient models for multi-dimensional signals processing. Traditional real-valued MLP models[2,3] have achieved state-of-the-art performance in dealing with gray-scale images. As for color images, it processed in an independent manner, in which the color image has been represented just as three gray-scale images to apply the real-valued MLP model for three times, and in this way, the correlation of the three color channels has been lost inevitably. Then, applications of complex numbers to neural networks have been extensively investigated[12,13] and have been well established in signal and image processing. Due to a growing number of studies concerning the use of quaternions in neural networks, Multilayer perceptron (MLP) models have been extended to quaternion domain[14−17]. The use of quaternion in MLP models yielded to a wide range of applications, such as color image compression[15], signal classification[16], etc. significantly owing to the powerful ability that quaternion has shown in the representation of color images. Simultaneously, the quaternion-valued MLP models produced large data redundancy and huge computation complexity, which mainly because of the existence of the real part in quaternion numbers and the noncommutativity of quaternion multiplication.

Addressing problems listed above, in this chapter, we propose an extended multilayer perceptron model based on reduced geometric algebra (RGA), called RGA-MLP, for multi-dimensional signals processing. Firstly, the proposed approach represents the input multi-dimensional signals as RGA multivectors[18], encoding the dimensions into components of RGA, to capture the fusion information of multiple dimensions. Secondly, all elements of RGA-MLP model including inputs, outputs, activation function and operators are extended to the RGA domain with commutative multiplication rules. Thirdly, we also present a RGA-version of back propagation (BP) algorithm to train the network. Finally, the performance of the proposed RGA-MLP model is evaluated by color images classification compared with the traditional real-valued MLP. The comparison experiments demonstrate that the proposed model outperforms the real one with respect to high classification accuracy and fast convergence rate.

The contributions of this chapter can be concluded as follows: Taking advantage of the RGA framework with commutative multiplication and none real part, which is bound to remove the data redundancy and simplify the structure of the networks extremely, the proposed RGA-MLP processes the multi-dimensional signals in a holistic manner without losing the relationship of different dimensions. Simultaneously, the weights updating and all operations in each layer have also been redefined based on the RGA theory in detail for the accomplishment of the networks. Specifically, it indeed shows optimal performance with lower training loss, higher test accuracy and faster convergence rate in contrast to the real one.

This chapter is organized into the following four sections. Section 9.2 introduces the basics reduced geometric algebra (RGA). Section 9.3 describes the basic structure of RGA-MLP model and the RGA-version of back propagation algorithm. In Section 9.4, the color images classification experiments are implemented to show the superiority of this approach and Section 9.5 gives the conclusions of the chapter.

9.2 Reduced Geometric Algebra (RGA)

Reduced geometric algebra (RGA) was introduced in Ref.[18], with commutative multiplication rules. The definition of RGA[18] can be given as follows.

$$\gamma_i = \frac{1}{2}\left(1 + e_i e_{n+i}\right), \quad i = 1, 2, \cdots, n \tag{9.1}$$

where e_i are the basis components of geometric algebra (GA), which can be referred to Ref.[19], and γ_1 denote the basis components of RGA, which obey the multiplication rules as follows

$$\gamma_i \gamma_j = \gamma_j \gamma_i, \quad i \neq j \tag{9.2}$$

At the same time, we define:

$$\gamma_i^2 = \gamma_i \gamma_i = \begin{cases} \gamma_{i+1}, & i = 1, 2, \cdots, n-1 \\ \gamma_1, & i = n \end{cases} \tag{9.3}$$

Take $n = 3$ for instance, $\gamma_1^2{=}\gamma_2\gamma_2^2{=}\gamma_3\gamma_3^2{=}\gamma_1$.

According to Eq.(9.2) and Eq.(9.3), the multiplication of i is evidently commutative. RGA is denoted as $\mathbb{G}_2^R$ and is the space generated by the collection of $\{\{\gamma_1, \gamma_2, \cdots, \gamma_n\}, \{\gamma_{ij} = \gamma_i \gamma_j, 1 \leqslant i \neq j \leqslant n\}\}$. In this chapter, we focus on the processing of color images which are three dimensions, the element x in $\mathbb{G}_2^R$ has the following form:

$$x = x^{(1)}\gamma_1 + x^{(2)}\gamma_2 + x^{(3)}\gamma_{12}, \quad x^{(1)}, x^{(2)}, x^{(3)} \in \mathbb{R} \tag{9.4}$$

The addition and subtraction in $\mathbb{G}_2^R$ are defined in the same manner as those in GA[19], here the multiplication operation is provided as follows. $\forall x, y \in \mathbb{G}_2^R$, suppose $x = x^{(1)}\gamma_1 + x^{(2)}\gamma_2 + x^{(3)}\gamma_{12}$ and then the multiplication in $y = y^{(1)}\gamma_1 + y^{(2)}\gamma_2 + y^{(3)}\gamma_{12}$ can be obtained as

$$\begin{aligned} xy &= \left(x^{(1)}\gamma_1 + x^{(2)}\gamma_2 + x^{(3)}\gamma_{12}\right)\left(y^{(1)}\gamma_1 + y^{(2)}\gamma_2 + y^{(3)}\gamma_{12}\right) \\ &= \left(x^{(1)}y^{(3)} + x^{(2)}y^{(2)} + x^{(3)}y^{(1)}\right)\gamma_1 + \left(x^{(1)}y^{(1)} + x^{(2)}y^{(3)} + x^{(3)}y^{(2)}\right)\gamma_2 \\ &\quad + \left(x^{(1)}y^{(2)} + x^{(2)}y^{(1)} + x^{(3)}y^{(3)}\right)\gamma_{12} \end{aligned} \tag{9.5}$$

where $\gamma_{12}\gamma_1 = \gamma_1\gamma_2\gamma_1 = \gamma_1\gamma_1\gamma_2 = \gamma_1^2\gamma_2 = \gamma_2\gamma_2 = \gamma_1$, similarly, $\gamma_{12}\gamma_2 = \gamma_2$. Moreover, $\gamma_{12}\gamma_{12} = \gamma_1\gamma_2\gamma_1\gamma_2 = \gamma_1\gamma_1\gamma_2\gamma_2 = \gamma_1^2\gamma_2^2 = \gamma_2\gamma_1 = \gamma_{12}$.

It can be seen form (9.5) that the results of the multiplication of $\mathbb{G}_2^R$, and y only contain γ_1, γ_2 and γ_{12} components, and no superfluous components are produced. Therefore, it overcomes the data redundancy embodied in the previous operations for color images based on quaternion.

The norm of the elements in $\mathbb{G}_2^R$ is defined as

$$\|x\| = a\gamma_1 + b\gamma_2 + c\gamma_{12} = |a+b+c| \triangleq \sqrt{a^2+b^2+c^2} \tag{9.6}$$

The conjugate of x is defined as

$$x^* = a'\gamma_1 + b'\gamma_2 + c'\gamma_{12} \tag{9.7}$$

Then

$$\begin{aligned} xx^* &= (a\gamma_1 + b\gamma_2 + c\gamma_{12})(a'\gamma_1 + b'\gamma_2 + c'\gamma_{12}) \\ &= (ca' + bb' + ac')\gamma_1 + (aa' + cb' + bc')\gamma_2 + (ba' + ab' + cc')\gamma_{12} \\ &= \|x\|^2 \end{aligned} \tag{9.8}$$

According to Eq.(9.6), the following equations are obtained

$$\begin{cases} ca' + bb' + ac' = 0 \\ aa' + cb' + bc' = 0 \\ ba' + ab' + cc' = \|x\|^2 \end{cases} \tag{9.9}$$

Solving the above equations yields the values of the individual components in Eq.(9.7), but not all the elements in $\mathbb{G}_2^R$ are conjugate.

Therefore, the inverse of element x in $\mathbb{G}_2^R$ is denoted as

$$x^{-1} = \frac{x^*}{\|x\|^2} \tag{9.10}$$

In RGA, multivectors, which are the extension of vectors to higher dimensions, are the basic elements. Any multivector $X \in G_2^R$ is so described by

$$X = X^{(1)}\gamma_1 + X^{(2)}\gamma_2 + X^{(3)}\gamma_{12} \tag{9.11}$$

where $X^{(1)}, X^{(2)}, X^{(3)} \in R(0)$.

9.3 Multilayer Perceptron Model Based on RGA for Color Images

9.3.1 RGA Neuron Model for Color Images

In this section, we utilize RGA to obtain an extension neuron model for multi-dimensional signals processing, in which all the operators can be extended to RGA domain. For each input color image F, each of its pixel can be represented in $\mathbb{G}_2^R$ as follows

$$f(x,y) = f^{(1)}(x,y)\gamma_1 + f^{(2)}(x,y)\gamma_2 + f^{(3)}(x,y)\gamma_{12} \tag{9.12}$$

Where $f^{(1)}(x,y)$, $f^{(2)}(x,y)$ and $f^{(3)}(x,y)$ are the R, G and B color components of color image pixel $f(x,y)$, respectively. Then input color image F can be denoted as a RGA multivector

$$F = F^{(1)}\gamma_1 + F^{(2)}\gamma_2 + F^{(3)}\gamma_{12} \tag{9.13}$$

The basic structure of the proposed RGA neuron model is shown in Fig.9.1.

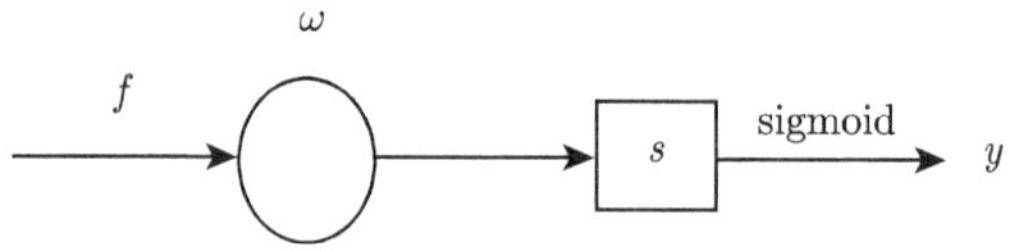

Fig. 9.1 The structure of RGA neuron model

For input $f \in \mathbb{G}_2^R$, the output $y \in \mathbb{G}_2^R$ can be formulated as follows

$$
\begin{aligned}
s &= \frac{f\omega}{\|\omega\|} - \theta \\
&= \frac{\left(f^{(1)}\gamma_1 + f^{(2)}\gamma_2 + f^{(3)}\gamma_{12}\right)\left(\omega^{(1)}\gamma_1 + \omega^{(2)}\gamma_2 + \omega^{(3)}\gamma_{12}\right)}{\sqrt{\left(\omega^{(1)}\right)^2 + \left(\omega^{(2)}\right)^2 + \left(\omega^{(3)}\right)^2}} - \theta \\
&= \frac{\begin{bmatrix} \left(f^{(1)}\omega^{(3)} + f^{(2)}\omega^2 + f^{(3)}\omega^{(1)}\right)\gamma_1 \\ \left(f^{(1)}\omega^{(1)} + f^{(2)}\omega^3 + f^{(3)}\omega^{(2)}\right)\gamma_2 \\ \left(f^{(1)}\omega^{(2)} + f^{(2)}\omega^1 + f^{(3)}\omega^{(3)}\right)\gamma_{12} \end{bmatrix}}{\sqrt{\left(\omega^{(1)}\right)^2 + \left(\omega^{(2)}\right)^2 + \left(\omega^{(3)}\right)^2}} - \theta
\end{aligned}
\tag{9.14}
$$

and

$$
y = g(s) \tag{9.15}
$$

where $\omega = \omega^{(1)}\gamma_1 + \omega^{(2)}\gamma_2 + \omega^{(3)}\gamma_{12} \in \mathbb{G}_2^R$ is the connection weight between neurons in the two layers, θ denotes the bias of the neuron and is also represented in $\mathbb{G}_2^R$. The function $g(\cdot)$ denotes a nonlinear activation function for neurons, here we adopt the sigmoid function and give its RGA version.

$$
g(s) = g(s^{(1)})\gamma_1 + g(s^{(2)})\gamma_2 + g(s^{(3)})\gamma_{12} \tag{9.16}
$$

Where $s = s^{(1)}\gamma_1 + s^{(2)}\gamma_2 + s^{(3)}\gamma_{12} \in G_2^R$, $s^{(1)}, s^{(2)}, s^{(3)} \in \mathbb{R}$ and $g(u) = \dfrac{1}{1+e^{-u}}$, $u \in \mathbb{R}$.

For instance, the γ_2 part of the sigmoid activation function output is shown in Fig.9.2 while γ_1 and γ_2 components of input are given as the axis.

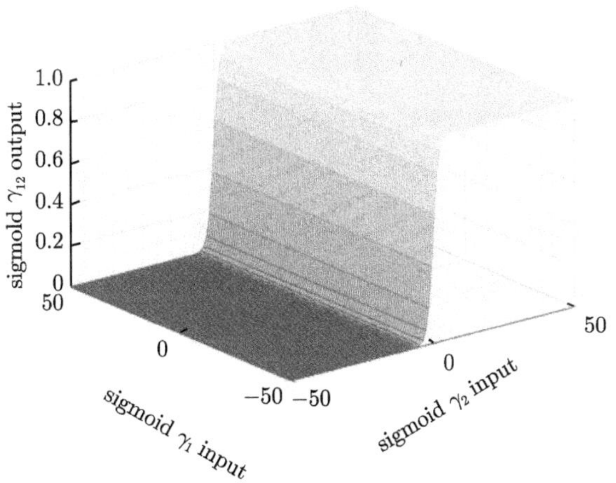

Fig. 9.2 RGA sigmoid function-γ_2 part output

As seen in Eq.(9.14), we can conclude that the input three-dimensional color image pixel $f \in \mathbb{G}_2^R$ is fully fused on different components of in $\mathbb{G}_2^R$ through multiplication rules RGA.

9.3.2 The Structure of RGA-MLP Model

Traditional MLP models for multi-dimensional signals treated each dimension signal as a real number and process it independently leading to some unsatisfying results. In this section we present an extension model of conventional multilayer perceptron (MLP) from real number to RGA domain to cater for multi-dimensional signals processing, and the RGA neurons as defined above form the basis of the proposed RGA-MLP model. Fig.9.3 depicts the structure of RGA-MLP model with one hidden layer.

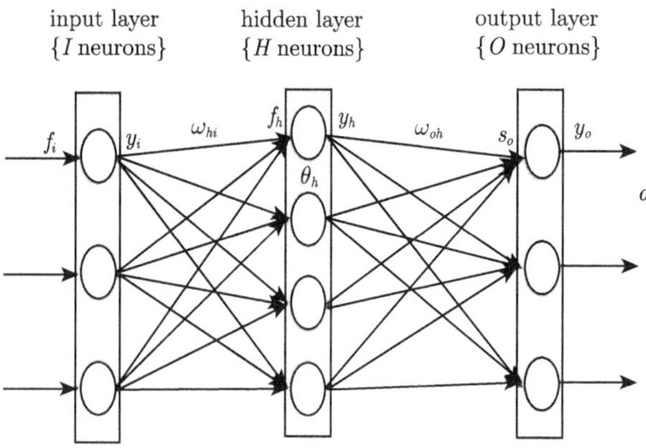

Fig. 9.3 The structure of RGA-MLP model with one hidden layer

The RGA-MLP model shown in Fig.9.3 consists of three layers, that is input layer, hidden layer and output layer, and the numbers of neurons are set to I, H and O, respectively. Specifically, $f_i \in \mathbb{G}_2^R, i = 1, 2, \cdots, I$ and $y_i \in \mathbb{G}_2^R, i = 1, 2, \cdots, I$ are the input and output of the i-th neuron in the input layer respectively. Similarly, $f_h \in \mathbb{G}_2^R, y_h \in \mathbb{G}_2^R, h = 1, 2, \cdots, H$ and $f_o \in \mathbb{G}_2^R, y_o \in \mathbb{G}_2^R, o = 1, 2, \cdots, O$ are the input and output of the h-th and o-th neuron in the hidden and output layers respectively. Moreover, the connection weights between the i-th and h-th neurons in input and hidden layers, between the h-th and o-th neurons in hidden and output layers, are denoted as $\omega_{hi} \in \mathbb{G}_2^R$ and $\omega_{oh} \in \mathbb{G}_2^R$. Significantly, the desired output for the o-th neuron in the output layer is $d_o \in \mathbb{G}_2^R$.

Then, refer to Eq.(9.14) and Eq.(9.15), it can be clearly obtained that

$$f_h = g\left(\sum_{i=1}^{I} y_i \omega_{hi} + \theta_h\right) \tag{9.17}$$

9.3.3 Learning Algorithm

In order to achieve the desired output $d_0 \in \mathbb{G}_2^R$, the connection weights are supposed to be modified by means of the so-called learning algorithms, here we adopt back-propagation (BP) algorithm and extended it to the RGA domain. Apropos of the RGA-MLP model with one hidden layer as given in Fig.9.3, we define the error E between the desired output and the target training data as

$$\begin{aligned} E &= E(y_1, \cdots, y_o, \cdots, y_O) \\ &= \frac{1}{2}\sum_{o=1}^{O}(y_o - d_o)^2 \\ &= \frac{1}{2}\sum_{o=1}^{O}\left(\left(y^{(1)}\gamma_1 + y^{(2)}\gamma_2 + y^{(3)}\gamma_{12}\right)_o - \left(d^{(1)}\gamma_1 + d^{(2)}\gamma_2 + d^{(3)}\gamma_{12}\right)_o\right)^2 \\ &= \frac{1}{2}\sum_{o=1}^{O}\left(\left(y_o^{(1)} - d_o^{(1)}\right)^2 + \left(y_o^{(2)} - d_o^{(2)}\right)^2 + \left(y_o^{(3)} - d_o^{3}\right)^2\right) \end{aligned} \tag{9.18}$$

where $y_o \in \mathbb{G}_2^R$ and $d_o \in \mathbb{G}_2^R$ are the output and the desired output of the o-th neuron in the output layer respectively.

In the drawback learning, the derivative of error function E with respect to any connection weights between two layers can be calculated as follows

$$\nabla E_{\omega_{ji}} = \nabla E_{\left(\omega^{(1)}\gamma_1 + \omega^{(2)}\gamma_2 + \omega^{(3)}\gamma_{12}\right)_{ji}} = \frac{\partial E}{\partial(\omega^{(1)})_{ji}}\gamma_1 + \frac{\partial E}{\partial(\omega^{(2)})_{ji}}\gamma_2 + \frac{\partial E}{\partial(\omega^{(3)})_{ji}}\gamma_{12} \tag{9.19}$$

Similarly, for the bias

$$\nabla E_{\theta_j} = \nabla E_{\left(\theta^{(1)}\gamma_1 + \theta^{(2)}\gamma_2 + \theta^{(3)}\gamma_{12}\right)_j} = \frac{\partial E}{\partial(\theta^{(1)})_j}\gamma_1 + \frac{\partial E}{\partial(\theta^{(2)})_j}\gamma_2 + \frac{\partial E}{\partial(\theta^{(3)})_j}\gamma_{12} \tag{9.20}$$

Then, the update of back-propagation for the RGA-MLP is defined as

$$\Delta\theta_j = -\eta\nabla E_{\theta_j} = -\eta\Psi_j \tag{9.21}$$

$$\Delta\omega_{ji} = -\eta \nabla E_{\omega_{ji}} = -\eta \Psi_j y_i^* \tag{9.22}$$

where * is set as the conjugation and η is a constant denoting the learning coefficient. In the case of the update of the o-th neuron in the output layer, ψ is given by

$$\begin{aligned}\Psi_o = &\left(g'(s)\left((y)_o - (d)_o\right)\right)^{(1)} \gamma_1 + \left(g'(s)\left((y)_o - (d)_o\right)\right)^{(2)} \gamma_2 \\ &+ \left(g'(s)\left((y)_o - (d)_o\right)\right)^{(3)} \gamma_{12}\end{aligned} \tag{9.23}$$

While ψ of the h-th neuron in the hidden layer is given by

$$\begin{aligned}\Psi_h = &\left(g'(s)\left(\left(\sum_o (\omega)_{ho}\right)^* (\Psi)_o\right)\right)^{(1)} \gamma_1 + \left(g'(s)\left(\left(\sum_o (\omega)_{ho}\right)^* (\Psi)_o\right)\right)^{(2)} \gamma_2 \\ &+ \left(g'(s)\left(\left(\sum_o (\omega)_{ho}\right)^* (\Psi)_o\right)\right)^{(3)} \gamma_{12}\end{aligned} \tag{9.24}$$

where

$$\begin{aligned}g'(s) &= g'(s^{(1)})\gamma_1 + g'(s^{(2)})\gamma_2 + g'(s^{(3)})\gamma_{12} \\ &= \frac{\partial g(s^{(1)})}{\partial s^{(1)}}\gamma_1 + \frac{\partial g(s^{(2)})}{\partial s^{(2)}}\gamma_2 + \frac{\partial g(s^{(3)})}{\partial s^{(3)}}\gamma_{12}\end{aligned} \tag{9.25}$$

9.4 Experiments and Analysis on Color Images Classification

In this section, we conduct color images classification experiments to illustrate the practical performance of the proposed RGA-MLP model compared with the traditional MLP model in the real domain, quantitatively and visually.

9.4.1 Experimental Setup

To implement color images classifications, the CIFAR-10 dataset, which consists of 60000 natural color images with 10 different classes and each of class contains 6000 images, is applied in this chapter. Among the 6000 images, the former 50000 images form training set and the rest 1000 images consists of testing set. All the color images in the dataset are composed of 32×32×3 (= 3072 pixels), whose values are normalized in the range of [0,1]. More details of the CIFAR-10 dataset can be found at http://www.cs.toronto.edu/kriz/cifar.html.

All the experiments in this chapter are performed by using MATLAB under the condition of Intel® Core^{TM} i5-6500 3.20GHz CPU and 8GB memory, Windows 7. We choose sigmoid as the activation function, and the RGA-version has been given in Fig.9.2. The training error and the test accuracy are evaluated on the three networks for comparison experiments.

9.4.2 Color Image Classification

In this section, color images classification experiments using traditional real-valued MLP model and the proposed RGA-MLP model are carried out based on the CIFAR-10 dataset. For the real-valued MLP model, each input color image is separated as three gray-scale images, ergo the number of neurons in the input layer is set to 3072 (=32×32×3), which is the same as the dimension of an input vector, and that in the output layer is 10 for the classification of 10 classes. In the proposed RGA-MLP model, the color image is represented as a RGA multivectors, thus the number of neurons in the input layer is 1024(=32×32) and that in the output layer in is set to 10. In this chapter, we select a three layer network with one hidden layer, in which the number of neurons seems to be indeterminate, and here we set that to 600 for the real-valued MLP model and 400 for the RGA-MLP model respectively. Then the structures of the two models can be denoted as 3072-600-10 and 1024-400-10.

We now turn to evaluate the influence of learning rate η, since it is important for the convergence of the proposed model. In the proposed RGA-MLP model, we consider the learning rateηas 0.01, 0.1, and 1 to explore the effect towards the rate of convergence, as shown in Fig.9.4. The network is trained for 600 iterations with a batch size of 10. As depicted in Fig.9.4, the proposed RGA-MLP can achieves the best convergence rate and the lowest training loss when the learning rate is 0.1. And the training loss becomes significantly stable after 10 epochs.

Then, we focus on the comparison experiments between the real-valued MLP model and RGA-MLP model with respect to the training error and the test accuracy, to exhibit the superiority of the proposed approach. The learning rate of the both models are fixed at 0.1 and the two models are trained for 1200 iterations while the batch size is set to 10. Fig.9.5 shows the training loss of the both models with respect to the number of training iterations.

More precisely, as shown in Fig.9.5, the proposed RGA-MLP model outperforms the real-valued MLP model with the faster rate of convergence under smaller number of iterations, that is to say, the proposed RGA-MLP model has the formidable ability in approaching a stable value more rapidly. Moreover, it can be clearly concluded that the proposed RGA-MLP achieves a more stable convergence trend and lower training errors while the real-valued MLP fluctuates greatly.

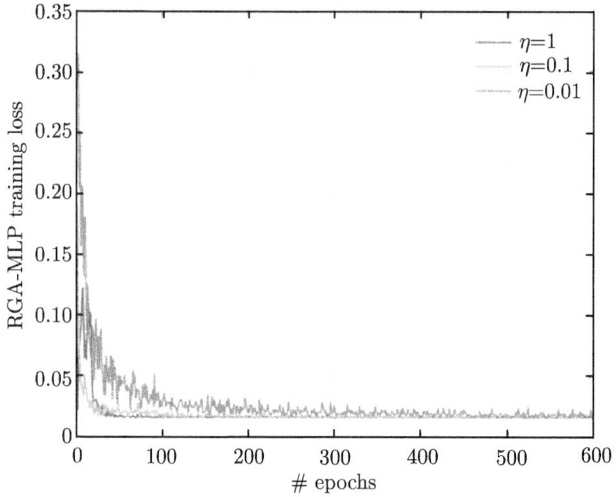

Fig. 9.4 The training loss curves achieved by the proposed RGA-MLP model with different learning rate η

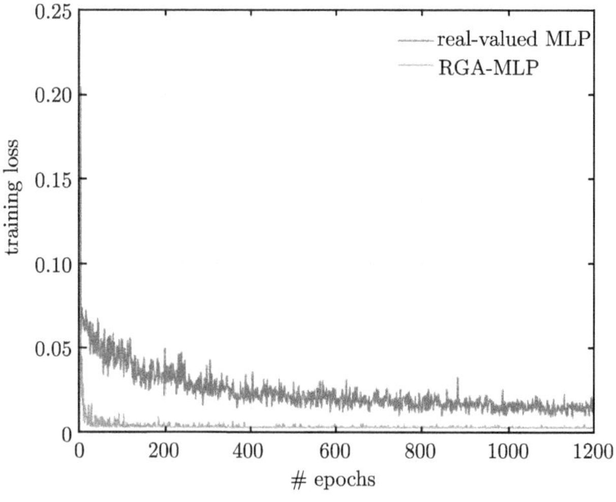

Fig. 9.5 The training loss curves achieved by the real-valued MLP and the proposed RGA-MLP models

For quantitatively comparison, training errors and test accuracies obtained by color images classification experiments performed by the real-valued MLP and the proposed RGA-MLP models are listed in Table 9.1. Expectedly, the proposed RGA-MLP exhibits lower training errors and higher test accuracies compared with the real-valued MLP. As suggested in Table 9.1, the test accuracy of classification yielded by the proposed RGA-MLP reaches approximately 84.8% while that of real-valued MLP is only 75.6%. To be more precise, the proposed approach applies RGA with commutative multiplication rules, which has the strong ability in capturing the relationship of color channels in color images when representing as RGA multivectors. Furthermore, a more simplified network and a powerful learning algorithm with higher learning ability are presented in the proposed RGA-MLP model, thereby superior performance is achieved compared with the traditional real-valued MLP model.

Table 9.1 Classification results of real-valued MLP and the proposed RGA-MLP models (%)

MLP models	structures	training errors	test accuracies
real-valued MLP	3072-600-10	1.76	75.6
RGA-MLP	1024-400-10	0.32	84.8

9.5 Conclusions

In this chapter, we have proposed an extended multilayer perceptron model and presented an error back-propagation algorithm for its learning scheme in the RGA domain, for color images processing. Taking advantage of RGA theory with commutative multiplication rules, the color images are represented as RGA multivectors, treating RGB color channels as a single unit instead of as separate components, and it is proved to be very effective. Specifically, the proposed approach, in which all elements and operators are extended into RGA domain, exploits strong ability in capturing the inherent color structures during the processing and significantly acts as an efficient and simplified network with formidable training ability. The superiority of the proposed model is then demonstrated compared with the traditional real-valued MLP model by experiments on color images classification. Our proposed model can achieve higher classification accuracy, faster convergence rate and lead to a more simplified network.

Applications of the proposed model to multi-dimensional vector data, such as multispectral images processing, predictions for three-dimensional protein structures, and controls of motion in three-dimensional space, are also promising and will be investigated in the future works.

References

[1] Perceptrons M L, Strapetz M, Fuchs R. Simultaneous evolution of structure and activation function types in generalized multi-layer perceptrons, WSES International Conference on Neural Network and Applications. CiteSeer, 2001.

[2] Dony R D, Haykin S. Neural network approaches to image compression. Proceedings of the IEEE 83, 1995: 288-303.

[3] Gallagher M, Downs T. Visualization of learning in multilayer perceptron networks using principal component analysis. IEEE Transactions on Systems Man Cybernetics Part B Cybernetics A Publication of the IEEE Systems Man Cybernetics Society, 2003, 33: 28-34.

[4] Chung F L, Wang S, Deng Z. Catsmlp: Toward a robust and interpretable multilayer perceptron with sigmoid activation functions. IEEE Transactions on Systems Man Cybernetics Part B Cybernetics A Publication of the IEEE Systems Man Cybernetics Society, 2006, 36(6): 1319.

[5] Gas B. Self-organizing multilayer perceptron. IEEE Transactions on Neural Networks, 2010, 21: 1766.

[6] Sivaram G S V S, Hermansky H. Sparse multilayer perceptron for phoneme recognition. IEEE Transactions on Audio Speech Language Processing, 2011, 20: 23-29.

[7] Rossi F, Conan-Guez B. Theoretical properties of functional multi-layer perceptrons with functional inputs. Neural Processing Letters, 2006, 23(1): 55-70.

[8] Rossi F, Conan-Guez B, Aicha E G. Clustering functional data with the som algorithm, Esann 2004, European Symposium on Artificial Neural Networks, Bruges, Belgium, April 28-30, 2004, Proceedings. DBLP, 2004: 305-312.

[9] Zhang D, Bai X L, Cai K Y. Extended neuro-fuzzy models of multilayer perceptrons. Fuzzy Sets Systems, 2004, 142: 221-242.

[10] Yang S S, Ho C L, Siu S. Sensitivity analysis of the split-complex valued multilayer

perceptron due to the errors of the i.i.d. inputs and weights. IEEE Transactions on Neural Networks, 2007, 18: 1280-1293.

[11] Grippo L, Manno A, Sciandrone M. Decomposition techniques for multilayer perceptron training. IEEE Transactions on Neural Networks Learning Systems, 2015, 27: 2146-2159.

[12] Kim T, Adal T. Approximation by fully complex multilayer perceptrons. Neural Computation, 2003, 15: 1641.

[13] Sivakumar P, Chithra N, Sivanandam S N, et al. Adaptive channel equalization using multiplicative neural network for Rayleigh faded channel. Journal of Computer Science, 2011, 7: 1646-1651.

[14] Teijiro I, Haruhiko N, Nobuyuki M. Quaternionic multilayer perceptron with local analyticity. Information, 2012, 3(4): 756-770.

[15] Luo L, Feng H, Ding L J. Color image compression based on quaternion neural network principal component analysis, International Conference on Multimedia Technology. IEEE, 2010: 1-4.

[16] Kusamichi H. A new scheme for color night vision by quaternion neural network. Proc. of the International Conference on Autonomous Robots and Agents, 2004.

[17] Buchholz S, Le B H N. Polarized signal classification by complex and quaternionic multi-layer perceptrons. International Journal of Neural Systems, 2008, 18: 75-85.

[18] Shen M M, Wang R, Cao W M. Joint sparse representation model for multichannel image based on reduced geometric algebra. IEEE Access. Image Process, 2018, 6: 24213-24223.

[19] Li Y S, Liu W M, Li X T, et al. GA-SIFT: A new scale invariant feature transform for multispectral image using geometric algebra. Information Sciences, 2014, 281: 559-572.

Chapter 10

RGA-CNNs: Convolutional Neural Networks Based on Reduced Geometric Algebra

10.1 Introduction

Recently, convolutional neural networks (CNNs) have exhibited high performance especially in object detection[1], face recognition[2] and image classification[3]. Since the thought of deep learning was introduced into neural networks motivated by the biological vision system, CNNs show great potential as active hotspots in artificial intelligence and computer vision currently[4−7]. As a classic kind of deep neural network, CNNs have several characteristics such as local connections, shared weights, pooling etc., which can reduce the number of training parameters and create some degree of invariance to shift, distortion and scale and show strong robustness and fault tolerance. Therefore, it is essential to train and optimize the network structure of CNNs. Based on these predominant characteristics, it has been shown to outperform the standard fully connected neural networks in a variety of signal and information processing tasks[8,9].

However, there are few researches on the CNNs model for multi-dimensional data, such as three-dimensional data which are typically presented as color images[10−14]. Traditional real-valued CNNs model[10−12] achieved state-of-the-art results in handling gray-scale images. As for color images, it just treats a color image as three gray-scale images to apply real-valued CNNs model independently, which potentially leads to the loss of fusion information of different color channels. Then, Guberman[13] proposed complex valued convolutional neural networks with complex valued input and parameters and it showed that a complex valued CNN can be seen as a restricted form of a larger real valued CNN, and as such it had the potential of mitigating the effects of over fitting. Kominami et al.[14] extended a part of CNNs to quaternion domain, called quaternion-valued convolutional neural networks (QCNNs), which efficiently handled problems of multiple inputs and multiple outputs. While the traditional neural networks that use real number algebra, has no ability to represent directly data in multi-dimensional spaces, especially three-dimensional signals such as color images. As all know that the way of efficiently dealing with color images by QCNNs model is to encode the three color channels into the three imaginary parts, due to the noncommutativity of multiplication in quaternion domain, it may inevitably produce a large plenty of data redundancy and a more complicated structure of network, moreover, it is also not conducive to the design of fast algorithm.

As above all, the existing CNNs models represent color image pixel as either a scalar[10−12] which loses some inherent color structures or a quaternion vector matrix[14] with high computational complexity. In this chapter, a novel convolutional neural networks based on RGA is presented, namely RGA-CNNs, which extend both the input color image pixels and convolution kernels as RGA multivectors to capture the intrinsic color structures, remove data redundancy and simplify the network. The theory of reduced geometric algebra (RGA) is introduced with commutative multiplication properties[15]. Utilizing the RGA theory, the model treats the input color image, neurons, convolution kernels, learning algorithm and all the related computations in the mathematic framework of RGA, where the basic operational rules, properties and convolution operations are defined. Thus the fusion information of different channels can be completely preserved and the complexity of the network can be consequently reduced. Furthermore, it demonstrates this approach on the challenging problem of classification tasks, and comparison experiments are conducted for color image to evaluate the feasibility of proposed model, which can preserve the inherent color structures and have higher learning ability where the learning cost of the network is lower, than the existing counterparts.

The rest of this chapter is organized as follows. Section 10.2 reviews the related work, including the basics of convolutional neural networks and geometric algebra. Section 10.3 describes the basic concepts of RGA and the RGA convolution. The structure of RGA-based CNNs model is proposed in Section 10.4. In Section 10.5, the classification experiments based on 3D geometric shapes and color images data sets are implemented to show the rationality and effectiveness of the proposed algorithm. Finally, the conclusion is given in Section 10.6.

10.2 Related Work

10.2.1 Convolutional Neural Network (CNN)

Convolutional neural network (CNN) is an efficient recognition algorithm which shows a wide range of applications in image processing, pattern recognition and other fields in recent years[11,12]. It has a simple structure, few training parameters and good adaptability and other advantages. The basic structure of CNN includes input, convolution, pooling, hidden, and output layers as shown in Fig. 10.1.

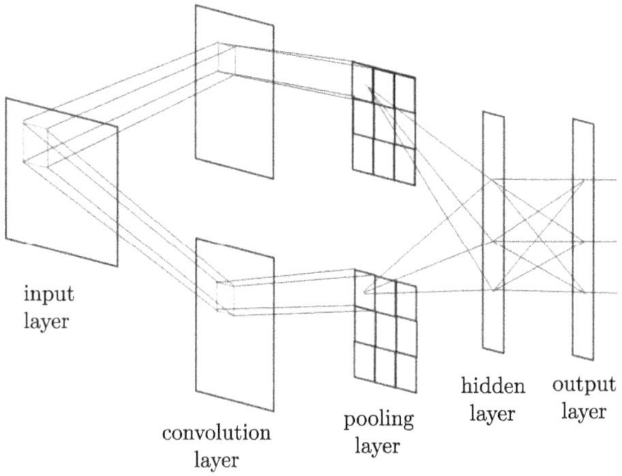

Fig. 10.1 The typical basic structure of CNNs

Evidently, fully connect hidden layers can usually be added before the output layer, and correspondingly after pooling layer. The convolution layer is used for extracting local features in images and then the pooling layer is applied to reduce the location sensitivity of the features which are extracted from convolution layer. Usually, the convolution layer and the pooling layer may repeat for several times in the structure of CNNs. In a convolution layer with input o, the j-th channel is given by

$$o \otimes J^{(j)} + b^{(j)} \tag{10.1}$$

where $\otimes$ is the convolution operation, and $\left\{J^{(j)}, b^{(j)}\right\}$ are the convolution kernels and bias terms, respectively.

Pooling is used to induce invariance to small translations, which is a characteristic of natural images. A pooling layer does so by splitting each input channel into patches, and replacing each patch with a single representative value in the output layer. Typical choices the maximal or average value, in max and average pooling, respectively. There exists instinct superiority that is, for large scale data, more layers and larger quantities of hidden elements are significantly necessary to be utilized. Then, by constructing such a structure, large-scale inputs can be reduced in size step by step and the characteristics becomes more obvious during learning process.

Recently, the research of CNNs theory has been extended from real number to hyper-complex number, typically presented as quaternion for three-dimensional data, such as color image. Since CNNs have proven to be very powerful in handling images, and quaternion numbers can produce meaningful representations in this domain, then the extension of real-valued convolutional neural networks to quaternion domain, namely quaternion-valued convolutional neural networks (QCNNs), have shown great potential in color image processing. Obviously, real-valued CNNs had achieved state-of-the-art performance when dealing with scalar data, typically as gray-scale image, and leading to some unsatisfying results with respect to color image in which the color image to be processed was usually separated into three independent color channels leading to the loss of relationship among three color channels. Then, QCNNs regard each color pixel as a pure quaternion and process multi-channel information in a parallel way, which mimic the human perception of a visual environment and the inherent color structure can be preserved completely.

However, a series of operations will produce a quaternion containing the real part, which will inevitably lead to the data redundancy. In addition, the quaternion multiplication is non-commutative, which is not conducive to design a simple network. Thus, QCNNs lead to a large number of data redundancy and a complicated network.

10.2.2 The Basics of Geometric Algebra

This section briefly reviews the basics of geometric algebra (GA), which was introduced by William K. Clifford and also called Clifford algebra[16–23]. Suppose $\mathbb{G}_n$ is n-dimensional GA with an orthonormal basis of vectors, which leads to a basis

$$\{1,\{e_i\},\{e_ie_j\},\cdots,\{e_1e_2\cdots e_n\}\} \tag{10.2}$$

Specifically, $\mathbb{G}_n$ can be represented by $\mathbb{G}_{p,q}$, where $n = p+q$ and p,q are the number of vectors with positive square and negative square in the basis of the space, respectively. That is

$$e_i^2=\begin{cases}1, & 1\leqslant i\leqslant p\\ -1, & p+1\leqslant i\leqslant n\end{cases} \tag{10.3}$$

In this chapter, we focus on $\mathbb{G}_n$, in which $q=0$. The geometric product of two basis is anti-commutative, and

$$e_ie_j=e_{ij}=-e_je_i=-e_{ji},\quad i,j=1,\cdots,n,\quad i\neq j \tag{10.4}$$

$$e_i^2=1,\quad i=1,\cdots,n \tag{10.5}$$

$$e_ie_{ij}=e_ie_ie_j=e_j,\quad i,j=1,\cdots,n,\quad i\neq j \tag{10.6}$$

For vectors ω and z of $\mathbb{G}_n$, the geometric product is defined as:

$$wz=w\cdot z+w\wedge z \tag{10.7}$$

where $w\cdot z$ denotes the inner product, $w\wedge z$ denotes the outer product. Since the e_i vectors are orthogonal, then $e_ie_j=e_i\cdot e_j+e_i\wedge e_j=e_i\wedge e_j$.

Take $\mathbb{G}_3$ for instance, the orthogonal bases are constructed by vectors with $2^3=8$ grades, which is given by: $\{1,\{e_1,e_2,e_3\},\{e_1e_2,e_1e_3,e_2e_3\},\{e_1e_2e_3\}\}$ and a simpler form: $\{1,e_1,e_2,e_3,e_{12},e_{13},e_{23},e_{123}\}$. For a 2^n-dimensional signal, only n basis are needed. Generally, the outer product of k vectors is called a k-blade. The number k is called the grade of the blade.

The complete mathematical correlation between two vectors has been provided by GA. As the extension of vectors to higher dimensions, multivectors are the basic components in GA. Any multivectors $M\in\mathbb{G}_n$ can be described by

$$M=E_0+\sum_{1\leqslant i\leqslant n}E_i(M)e_i+\sum_{1\leqslant i<j\leqslant n}E_{ij}(M)e_{ij}+\cdots+E_{1\cdots n}(M)e_{1\cdots n} \tag{10.8}$$

where $E(M)\in\mathbb{R}$.

10.3 Reduced Geometric Algebra (RGA)

In terms of the noncommutativity of multiplication given by GA, which results in the high algorithm complexity, this chapter introduces a geometric algebra theory with commutative properties [15].

10.3.1 The Basics of RGA

The definition of the reduced geometric algebra (RGA)[15] is given as follows:

$$f(x,y) = \sum_{i=1}^{n} f_i(x,y)\, e_i \tag{10.9}$$

$$\varepsilon_i = \frac{1}{2}(1 + e_i e_{n+i}) \in \mathbb{G}_n^R, \quad i = 1, 2, \cdots, n \tag{10.10}$$

According to Eq.(10.4), the geometric product of and can be obtained by:

$$\varepsilon_i \varepsilon_j = \varepsilon_j \varepsilon_i, \quad i \neq j \tag{10.11}$$

Moreover, we define:

$$\varepsilon_i^2 = \varepsilon_i \varepsilon_j = \begin{cases} \varepsilon_{i+1}, & i = 1, 2, \cdots, n-1 \\ \varepsilon_1, & i = n \end{cases} \tag{10.12}$$

Take $n = 3$ for instance, $\varepsilon_1^2 = \varepsilon_2, \varepsilon_2^2 = \varepsilon_3, \varepsilon_3^2 = \varepsilon_1$. Thinking together Eq.(10.10) and Eq.(10.11), it is clearly that the multiplication of ε_i is commutative. Specifically, RGA is denoted as $\mathbb{G}_n^R$ and can be seen as the space that is generated by the collection of $\{\{\varepsilon_1, \varepsilon_2, \cdots, \varepsilon_n\}, \{\varepsilon_{ij} = \varepsilon_i \varepsilon_j, 1 \leqslant i \neq j \leqslant n\}\}$. The element k in $\mathbb{G}_2^R$ has the following form:

$$\varphi(G_{n+2} \otimes f \otimes G_{n+2}) = \sum_{j=1}^{n} \left(\frac{1}{M^2 N^2} \sum_{l=0}^{M-1} \sum_{s=0}^{N-1} \sum_{p=0}^{M-1} \sum_{q=0}^{N-1} \sum_{i=1}^{n} g_i f_i g_j' e_j \right) \tag{10.13}$$

$$k = a^1 \varepsilon_1 + a^2 \varepsilon_2 + a^3 \varepsilon_{12}, \quad a^1, a^2, a^3 \in \mathbb{R} \tag{10.14}$$

The addition and subtraction operations in $\mathbb{G}_2^R$ are almost the same as GA, here we present the multiplication operation only. $\forall k, l \in \mathbb{G}_2^R$, suppose $k = a^1\varepsilon_1 + a^2\varepsilon_2 + a^3\varepsilon_{12}$ and $l = b^1\varepsilon_1 + b^2\varepsilon_2 + b^3\varepsilon_{12}$, then the multiplication in $\mathbb{G}_2^R$ are given as

$$\begin{aligned} kl &= \left(a^1\varepsilon_1 + a^2\varepsilon_2 + a^3\varepsilon_{12}\right)\left(b^1\varepsilon_1 + b^2\varepsilon_2 + b^3\varepsilon_{12}\right) \\ &= \left(a^1b^3 + a^2b^2 + a^3b^1\right)\varepsilon_1 + \left(a^1b^1 + a^2b^3 + a^3b^2\right)\varepsilon_2 \end{aligned}$$

$$+\left(a^1b^2+a^2b^1+a^3b^3\right)\varepsilon_{12} \tag{10.15}$$

where $\varepsilon_{12}\varepsilon_1=\varepsilon_1\varepsilon_2\varepsilon_1=\varepsilon_1\varepsilon_1\varepsilon_2=\varepsilon_1^2\varepsilon_2=\varepsilon_2\varepsilon_2=\varepsilon_1$, similarly, $\varepsilon_{12}\varepsilon_2=\varepsilon_2$. Moreover, $\varepsilon_{12}\varepsilon_{12}=\varepsilon_1\varepsilon_2\varepsilon_1\varepsilon_2=\varepsilon_1\varepsilon_1\varepsilon_2\varepsilon_2=\varepsilon_1^2\varepsilon_2^2=\varepsilon_2\varepsilon_1=\varepsilon_{12}$.

It can be seen form (10.13) that no superfluous components are produced in the results of the multiplication of k and l, which only contain ε_1, ε_2 and ε_{12} components. In this way, it successfully overcomes the data redundancy existed in the operations for color image based on quaternion.

We then define the norm of the elements in $\mathbb{G}_2^R$ as

$$\|k\|=a\varepsilon_1+b\varepsilon_2+c\varepsilon_{12}=|a+b+c|=\sqrt{a^2+b^2+c^2} \tag{10.16}$$

The conjugate of k is defined as

$$k^*=a'\varepsilon_1+b'\varepsilon_2+c'\varepsilon_{12} \tag{10.17}$$

Then

$$\begin{aligned} kk^* &= \left(a\varepsilon_1+b\varepsilon_2+c\varepsilon_{12}\right)\left(a'\varepsilon_1+b'\varepsilon_2+c'\varepsilon_{12}\right) \\ &= \left(ca'+bb'+ac'\right)\varepsilon_1+\left(aa'+cb'+bc'\right)\varepsilon_2 \\ &\quad+\left(ba'+ab'+cc'\right)\varepsilon_{12}=\|k\|^2 \end{aligned} \tag{10.18}$$

According to Eq.(10.16), the following equations are obtained

$$\begin{cases} ca'+bb'+ac'=0 \\ aa'+cb'+bc'=0 \\ ba'+ab'+cc'=\|k\|^2 \end{cases} \tag{10.19}$$

When solving the equations in Eq.(10.17), the values of the individual components in Eq.(10.15) are correspondingly yielded, but not all the elements of $\mathbb{G}_2^R$ are conjugate.

Thus, the inverse of element k in $\mathbb{G}_2^R$ can be defined as

$$k^{-1}=\frac{k^*}{\|k\|^2} \tag{10.20}$$

In RGA, multivectors, which are the extension of vectors to higher dimensions, are the basic units. Each multivector $K\in\mathbb{G}_2^R$ is so described by

$$K=K^1\varepsilon_1+K^2\varepsilon_2+K^3\varepsilon_{12} \tag{10.21}$$

where $K^1,K^2,K^3\in\mathbb{R}$.

10.3.2 Convolution in RGA

Since a convolution in image space is stated equally to multiplication in frequency space, plays a crucial role in signal processing. In this section, the convolution theorem using the theory of RGA is proposed for the first time.

Refer to the definition of convolution for multivectors in GA space[21], we give the convolution of two multivectors in $\mathbb{G}_2^R$.

Now, let $K = K^1\varepsilon_1 + K^2\varepsilon_2 + K^3\varepsilon_{12}$ and $L = L^1\varepsilon_1 + L^2\varepsilon_2 + L^3\varepsilon_{12}$ be a multivector field a multivector-valued filter, respectively. The RGA convolution is defined as

$$\boldsymbol{c}(x, y) = \iint_{\mathbb{G}_2^R} K(x, y)L(x - u, y - v)\mathrm{d}u\mathrm{d}v \tag{10.22}$$

where (x, y) indicates the coordinate. For discrete multivector space, the convolution has to be discretized and can be deduced as follows

$$\boldsymbol{c}(x, y) = \frac{1}{MN}\sum_{i=0}^{\infty}\sum_{j=0}^{\infty} K(x, y)L(x - i, y - j) \tag{10.23}$$

Clearly, each color image pixel is represented as a RGA multivector, the convolution can be performed referring to Eq.(10.21). Since the multiplication in $\mathbb{G}_2^R$ is commutative ($KL = LK$), the right-side and the left-side convolutions can be quite the same.

Suppose $f(x, y) \in \mathbb{G}_2^R$ and $g(x, y) \in \mathbb{G}_2^R$ are a color image and a filter in RGA space respectively. Consequently, we propose the definition of convolution in RGA for color image as follows

$$f(x, y) \otimes g(x, y) = \frac{1}{MN}\sum_{i=0}^{M-1}\sum_{j=0}^{N-1} f(x - i, y - j)g(i, j) \tag{10.24}$$

where M and N denote the size of convolution window.

10.4 Convolutional Neural Networks Based on Reduced Geometric Algebra

10.4.1 The Structure of RGA-CNNs

Since existing CNNs treat each color image pixel as a scalar with scalar input and scalar convolution kernels which loses some color structures, or a pure quaternion with high data redundancy and a complicated network.

In this chapter, we propose a novel convolutional neural networks model based on RGA, namely RGA-CNNs, which extend both the input color image pixels and convolution kernels as RGA multivectors, and all the operations are extended into the RGA domain, to capture the inherent color structures, simplify the network and remove data redundancy. The basic structure of the proposed RGA-CNNs, in which each representation and operation are based on RGA, is shown in Fig. 10.2, including input layer, convolution layer, pooling layer, multilayer perceptron (MLP) and output layer.

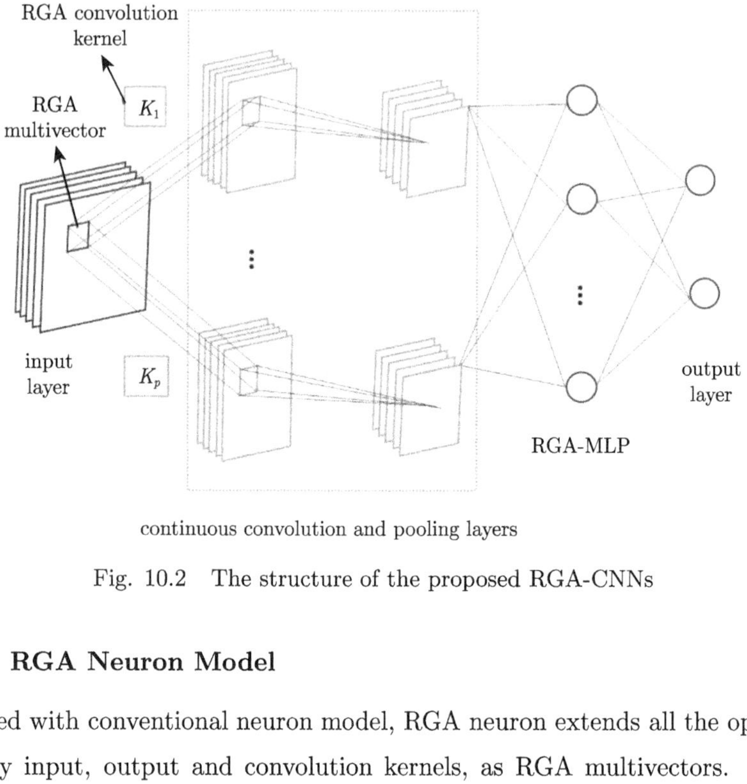

Fig. 10.2 The structure of the proposed RGA-CNNs

10.4.2 RGA Neuron Model

Compared with conventional neuron model, RGA neuron extends all the operators, especially input, output and convolution kernels, as RGA multivectors. Fig.10.3 shows the structure of the proposed RGA neuron model. For the input RGA signal $x \in \mathbb{G}_2^R$, the output $y \in \mathbb{G}_2^R$ is represented as

$$\text{recall} = \frac{\text{number of correct-positives}}{\text{total number of positives}} \tag{10.25}$$

$$y = f(s) \tag{10.26}$$

and

$$\text{precision} = \frac{\text{number of correct-positives}}{\text{total number of matches(correct or false)}} \tag{10.27}$$

$$\begin{aligned} s &= \frac{\omega x}{\|\omega\|} - \varphi = \frac{(\omega^1\varepsilon_1 + \omega^2\varepsilon_2 + \omega^3\varepsilon_{12})(x^1\varepsilon_1 + x^2\varepsilon_2 + x^3\varepsilon_{12})}{\sqrt{(\omega^1)^2 + (\omega^2)^2 + (\omega^3)^2}} - \varphi \\ &= \frac{\begin{bmatrix} (\omega^1x^3 + \omega^2x^2 + \omega^3x^1)\varepsilon_1 \\ (\omega^1x^1 + \omega^2x^3 + \omega^3x^2)\varepsilon_2 \\ (\omega^1x^2 + \omega^2x^1 + \omega^3x^3)\varepsilon_{12} \end{bmatrix}}{\sqrt{(\omega^1)^2 + (\omega^2)^2 + (\omega^3)^2}} - \varphi \end{aligned} \tag{10.28}$$

where $\omega \in \mathbb{G}_2^R$ denotes the weight of the connections between neurons, φ is the internal state of the RGA neuron and also called bias.

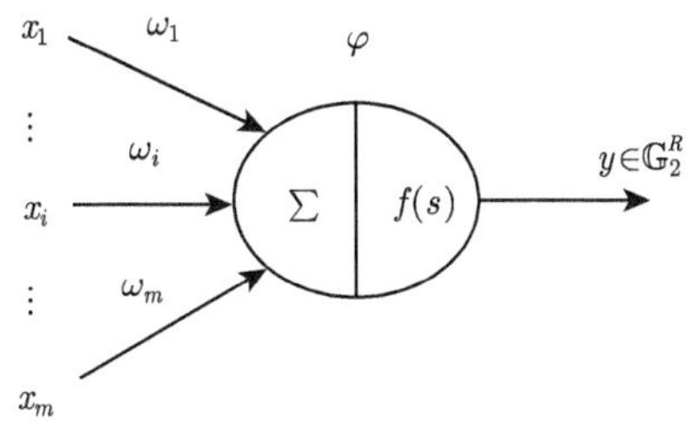

Fig. 10.3 The structure of RGA neuron model

The output of the neuron is determined by the output of the activation function $f(\cdot)$, which is defined by

$$\text{precision} = \frac{\text{number of correct-positives}}{\text{total number of matches(correct or false)}} \tag{10.29}$$

$$f(s) = f(s^1)\varepsilon_1 + f(s^2)\varepsilon_2 + f(s^3)\varepsilon_{12} \tag{10.30}$$

where $s = s^1\varepsilon_1 + s^2\varepsilon_2 + s^3\varepsilon_{12} \in \mathbb{G}_2^R$, $s^1, s^2, s^3 \in \mathbb{R}$ and $f(u) = \dfrac{1}{1+e^{-u}}, u \in \mathbb{R}$.

Obviously, a left-side (right-side) weight association neuron is generally inferior to a spinor one in terms of learning ability in some tasks. However, the computational complexity of the network with spinor neuron is more complicated than that of the single-side one due to the two-side weight association. Fortunately, RGA is commutative ($\omega x = x\omega$), that is to say the single-side weight association neuron can achieve almost the same learning ability as spinor neuron while removing the data redundancy and simplifying the network.

10.4.3 RGA Multilayer Perceptron (RGA-MLP) and Its Learning Algorithm

In this section we extend conventional multilayer perceptron (MLP) from real number to RGA, in which all the signals, weights and activation values are substituted for RGA numbers taken from a given RGA, to accommodate color image processing. Fig. 10.4 shows the basic structure of RGA-MLP with 3 fully connect hidden layers.

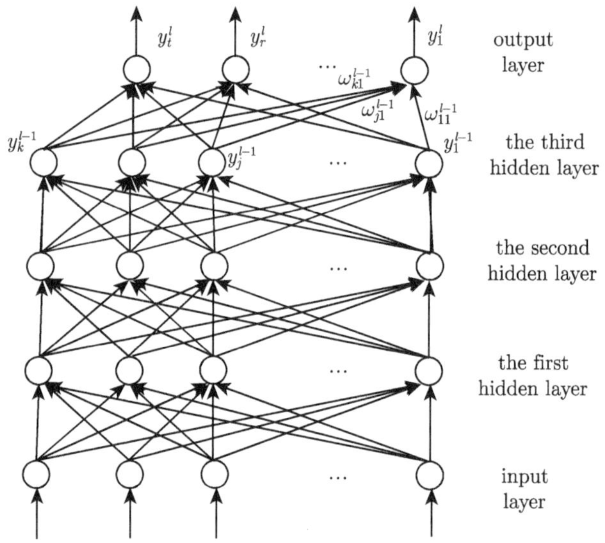

Fig. 10.4 The structure of RGA-MLP containing 3 hidden layers

In Fig.10.4, the input and output signals of each neuron, the weights between two neurons of adjacent two layers are all RGA multivectors, that is $y, \omega \in \mathbb{G}_2^R$. y_j^{l-1} denotes the output of the neuron j of the $(l-1)$-th layer and ω_{jr}^{l-1} means the connection weight between neuron j in the $(l-1)$-th layer and neuron r-th layer, that is the output layer. Suppose $x_i^l = \left(x^1\varepsilon_1 + x^2\varepsilon_2 + x^3\varepsilon_{12}\right)_i^l \in \mathbb{G}_2^R$ is the input of neuron i of the l-th layer in RGA-MLP, $y_i^l \in \mathbb{G}_2^R$ and $\varphi_i^l \in \mathbb{G}_2^R$ are output and bias of this neuron. $\omega_{ji}^{l-1} = \left(\omega^1\varepsilon_1 + \omega^2\varepsilon_2 + \omega^3\varepsilon_{12}\right)_{ji}^{l-1} \in \mathbb{G}_2^R$ denotes the weight between neuron j in the $(l-1)$-th layer and neuron i in the l-th layer, then

$$\text{precision} = \frac{\text{number of correct-positives}}{\text{total number of matches(correct or false)}} \tag{10.31}$$

$$x_i^l = \varphi_i^l + \sum_{j=1}^{k} \omega_{ji}^{l-1} y_j^{l-1} \tag{10.32}$$

$$\text{precision} = \frac{\text{number of correct-positives}}{\text{total number of matches(correct or false)}} \tag{10.33}$$

$$y_i^l = f(x_i^l) \tag{10.34}$$

As a normally and effective updating algorithm, back propagation (BP) is utilized to minimize the output error, which can be calculated by outputs of neurons in the output layer and the desired output signals. In this chapter, BP algorithm has been extended to the RGA domain and then adopted to train the proposed RGA-CNNs scheme.

The process of our proposed RGA-CNNs can be summarized as follows.

In this chapter, we extend conventional convolution kernel in real domain to RGA domain. As shown in Fig. 10.5, the input matrix and the convolution kernel are represented as RGA multivectors $A \in \left(\mathbb{G}_2^R\right)^{M\times M}$ and $K \in \left(\mathbb{G}_2^R\right)^{N\times N}$, respectively.

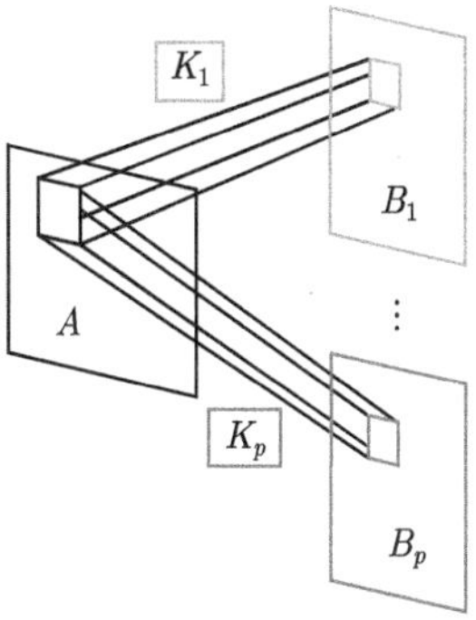

Fig. 10.5 The convolution layer of RGA-CNNs

In the convolution layer, the output $B \in \left(\mathbb{G}_2^R\right)^{(M-N+1)\times(M-N+1)}$ is generated by convolution of input with convolution kernel K as follows

$$\text{precision} = \frac{\text{number of correct-positives}}{\text{total number of matches(correct or false)}} \tag{10.35}$$

$$b = f\left(\sum \frac{ka}{|k|} + \varphi\right) \tag{10.36}$$

where k, a, b stand for each pixel of k, a, b respectively, and $b = f(\cdot)$ has been defined by Eq.(10.3). Then a feature matrix B is obtained and so called feature map. Since the convolution layer is composed of several feature maps, that is to say, p feature maps $B_1, B_2, \cdots, B_p$ can be generated through convolution of input A with p RGA convolution kernels $K_1, K_2, \cdots, K_p$.

The output B is fed to the pooling layer, which aims to reduce the position sensitivity of the features extracted by convolution layer. Here, we extend the conventional max pooling method in the real domain to RGA domain, and specifically it is denoted as

$$\text{precision} = \frac{\text{number of correct-positives}}{\text{total number of matches(correct or false)}} \tag{10.37}$$

$$c_{ij} = \max_{(k,l)\in C_{ij}} b_{kl} \tag{10.38}$$

where C_{ij} is the local patch in the output of pooling layer C, c_{ij} is the pixel of C_{ij}, and b is the pixel of the output of convolution layer B.

Generally speaking, after several convolution and pooling layers, the final outputs are offered to the RGA-MLP. The proposed CNNs model, in which the neural and convolution operations, activation functions, pooling and learning algorithm are all performed under the framework of RGA, color image are processed in a holistic manner without losing the interrelationship of the three channels and reducing the computational complexity due to the commutative multiplication rule.

10.5 Classification Experiments and Analysis

The convolutional neural network structures described above are applicable to multi-dimensional application domains, in which a functional dependency exists among the dimensions. One of the attractive and challenging applications able to validate the superiority and advantages of the proposed RGA-CNNs is that of classification. In this section, we perform both 3D geometrical shapes and color image classification experiments to evaluate the practical performance and computational efficiency of the proposed RGA-CNNs model compared with the existing real-valued CNNs[10] and QCNNs[14] models, quantitatively and visually.

10.5.1 Data Sets

1. 3D Geometrical Shapes Data Sets

To form the 3D geometrical shapes data sets, we generate 4 classes of 3D geometrical shapes, in which 4000 patterns are contained in each of class. In terms of our 3D geometrical shapes data sets, different shapes belong to different classes and different patterns of the same class are generated by the rotation of the first pattern with different angles, around the origin and by the translation with different vectors. Fig. 10.6 shows the example 3D geometrical shapes with 4 classes and 4 patterns for each of class, used in the classification experiments.

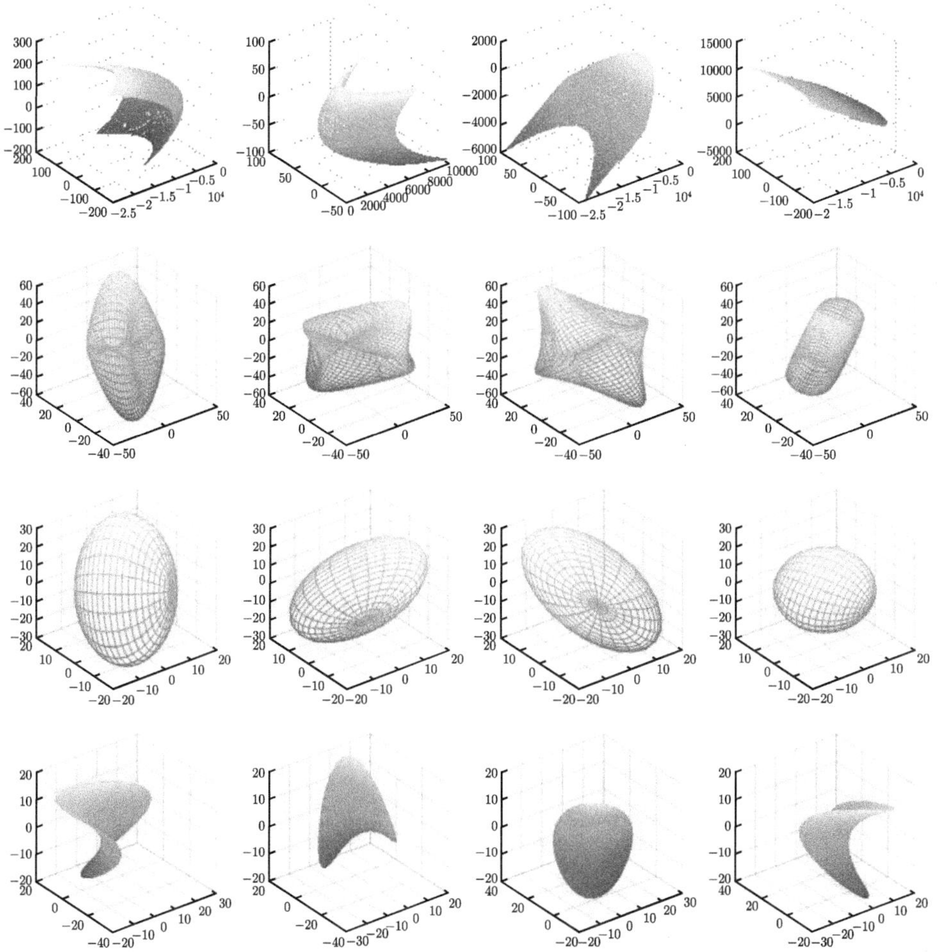

Fig. 10.6 Four classes of plane curves

2. Color Image Data Set

In this chapter, we include the CIFAR-10 dataset for color image classification experiments. The CIFAR-10 dataset comprises totally 60000 natural color images with 10 different classes and 6000 images for each of class, in which 50000 images are acted as the training set and the rest 1000 images are applied as the testing set. All the images in the dataset consists of 32×32 (= 1024 pixels), each of which has its color values represented by 24 bits (red, green and blue components are represented by 8 bits). More details of the CIFAR-10 dataset can be found at http://www.cs.toronto.edu/~kriz/cifar.html.

10.5.2 Experimental Setup

The architectures of real-valued CNNs, QCNNs and the proposed RGA-CNNs are mainly similar and composed of an input layer, a convolution layer, a pooling layer, a MLP and an output layer, respectively. The number of convolution kernels in the convolution layer is set to 12 with the size of 5×5 and pooling size is 2×2. Specifically, the input data of size 32×32 becomes 28×28 after convolution with a 5×5 convolution kernel. After max pooling of 4×4, output of pooling layer becomes 6×6. For k class problem, output layer has k neurons. Outputs of pooling layer are converted to quaternion or RGA numbers for QCNNs and RGA-CNNs, respectively. We initialize the learning rate η to 0.1 for the three networks. All the experiments in this chapter are performed by using MATLAB on a computer with Intel® Core™ i5-3470 3.20GHz CPU and 4GB memory. We choose sigmoid as the activation function, and max pooling by magnitude. The training error and the test accuracy are evaluated on the three networks for comparison experiments.

10.5.3 3D Geometrical Shapes Classification

Based on the 3D geometrical shapes database described above, we focus on analyzing the influence on the network performance, including the network depth, the convolution kernel size and the convolution kernel number. We choose 16000 3D geometrical shapes from 4 classes (where 3000 shapes and 1000 of each class are selected to construct the training dataset and the test dataset, respectively).

1. The Size of Convolution Kernels

Since the network structure of convolution model is more complex, which leads to relatively long training time, how to find the most suitable network structure with reasonable size of convolution kernels and less network loss value has become the main consideration in this section.

In order to explore the relationship between the performance of the proposed RGA-CNNs and the size of convolution kernels in the convolution layer, we conduct 3D geometrical shapes classification experiments in which the size of convolution kernels are set to 3×3, 5×5, 7×7, 9×9 and 11×11. The training loss achieved by the proposed RGA-CNNs with different size of convolution kernels is depicted in Fig.10.7. And Table 10.1 shows the average training time of QCNNs and the proposed RGA-CNNs with different size of kernels. It is clear that with the continuously increasing size of convolution kernels, the training loss values tend to be slightly lower and have little influence on the rate of converge in 3D geometrical shapes classification experiments.

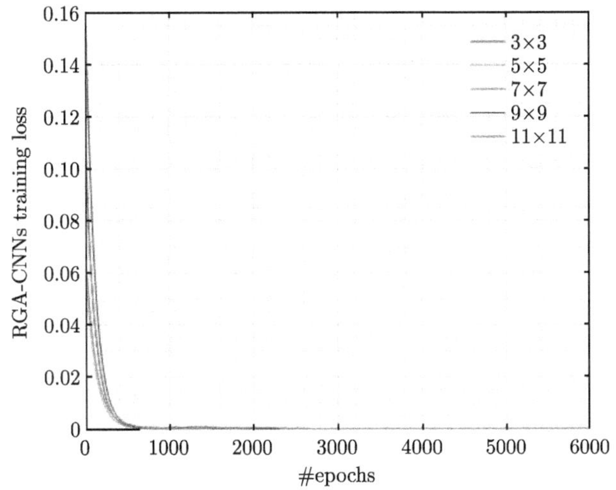

Fig. 10.7 The training loss curves achieved by the proposed RGA-CNNs with different size of convolution kernels

Table 10.1 Average training time (s) of QCNNs and the proposed RGA-CNNs with different size of convolution kernels

the size of convolution kernels	3×3	5×5	7×7	9×9	11×11
QCNNs[14]	2509.9803	2773.0203	3110.7038	3519.2984	3921.6121
RGA-CNNs	1527.4333	1677.9276	1828.0485	1999.1379	2176.4174

However, much more pixel data are correspondingly read while processing the whole feature image with the increasing size of convolution kernels, which inevitably leads to higher training time of the network, and the lager the size of convolution kernel is, the higher training time it takes. And it is obvious that the average training time of the proposed RGA-CNNs is much lower than QCNNs, for the reason that RGA is commutative while quaternion is noncommutative. Thus, combine the effects of training errors and training times, we choose 5×5 as the most reasonable size of convolution kernels.

2. Comparisons of Real-valued CNNs, QCNNs and the Proposed RGA-CNNs

The training loss curves achieved by real-valued CNNs and the proposed RGA-CNNs based on 3D geometrical shapes database are depicted in Fig.10.8 for comparison, in which the convolution kernels are all set to 5×5. We trained both the networks for 6000 iterations with a batch size of 10.

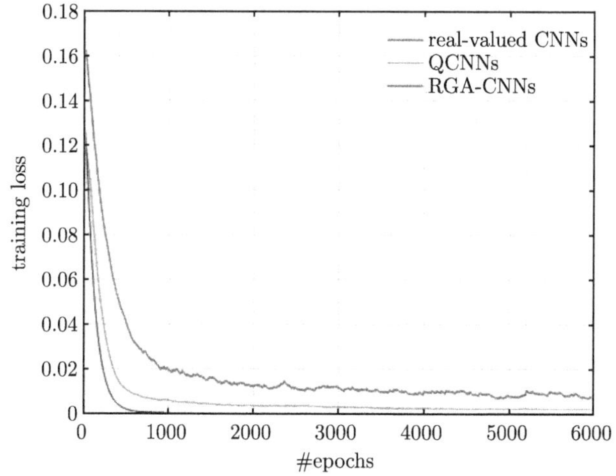

Fig. 10.8 The training loss curves achieved by the real-valued CNNs, QCNNs and the proposed RGA-CNNs

As shown in Fig.10.8, it is clearly that the proposed RGA-CNNs achieve faster and more stable converge rate compared to real-valued CNNs and QCNNs. Furthermore, under the same iterations, the traditional real-valued CNNs shows a little bit higher training errors in contrast to QCNNs, and then the training errors achieved by the proposed RGA-CNNs achieve are lower than QCNNs.

To give a more precise comparison between the proposed RGA-CNNs, QCNNs and the traditional real-valued CNNs, the training errors with respect to the number of training iterations, and the results are listed in Table 10.2.

Table 10.2 The training errors of the proposed RGA-CNNs, QCNNs and real-valued CNNs with respect to training iterations

iterations	1	400	1000	2000	4000	6000
real-valued CNNs[10]	0.1626	0.0507	0.0193	0.0124	0.0093	0.0069
QCNNs[14]	0.1192	0.0146	0.0060	0.0036	0.0024	0.0020
RGA-CNNs	0.1261	0.0038	2.4937×10^{-4}	2.4937×10^{-4}	2.4937×10^{-4}	2.4937×10^{-4}

As can be indicated in Table 10.2, when the number of training iterations go up from 1 to 400, the corresponding training errors of the traditional real-valued CNNs go down from 0.1626 to 0.0507 with reduction of nearly 0.1119, and the training errors of QCNNs decrease from 0.1192 to 0.0146 with reduction of nearly 0.1046, while that of the proposed RGA-CNNs decrease from 0.1261 to 0.0038 with reduction of nearly 0.1223. That is to say, with the first 400 iterations of training, the training errors of the proposed RGA-CNNs exhibit an extremely steeper downward trend compared with real-valued CNNs and QCNNs, commonly shown as a faster convergence rate. And as the iterations range from 1000 to 6000, the real-valued CNNs and QCNNs show approximately 0.0124 and 0.0040 decrease of training errors while the proposed RGA-CNNs reveal a nearly 0.0002 decrease, usually presented to be more stable.

That is, the proposed RGA-CNNs outperform the existing real-valued CNNs and QCNNs in terms of training loss due to the powerful and efficiently tool provided by RGA framework, in which all the training and test data in 3D geometrical shapes database are represented as RGA multivectors and can then be processed in a holistic way. Consequently, the correlation of different dimensions can be successfully retained and a simpler network can be obtained owing to the communicative multiplication and none real part of RGA.

10.5.4 Color Image Classification

We next investigate the performance for the color image classification task based on the CIFAR-10 dataset. In this experiment, 4 classes of color images in CIFAR-10 are used, 6000 color images of each class in training database are selected to form the training set and the test set is composed of 2000 color images of each class in test database. The networks are trained so that the original input should be reconstructed at the output layer.

1. Learning Rate

Since the learning rate is important for the convergence of the proposed network. We conducted an experiment to discuss the effect of the learning rate η on the performance of the proposed RGA-CNNs. We correspondingly set the learning rate η to 0.0001, 0.001, and 0.01.

As indicated in Fig.10.9, the proposed RGA-CNNs can achieves good convergence and the lowest training loss with different learning rates. When the learning rate η is set to 0.01, the proposed RGA-CNNs can achieve the best convergence. It is easily indicated that by adjusting the learning rate, the approximately optimal solution with respect to the optimization problem of training loss function can be obtained.

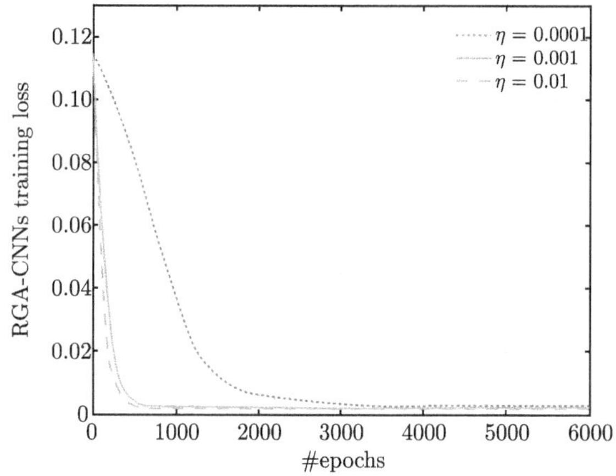

Fig. 10.9 The training loss curves achieved by the proposed RGA-CNNs with different learning rate η

2. Comparisons of Real-valued CNNs, QCNNs and the Proposed RGA-CNNs

We evaluate the training error and the test accuracy on real-valued CNNs and the proposed RGA-CNNs for comparison experiments. For real-valued CNNs the learning rate is fixed at 0.1, for QCNNs and RGA-CNNs, the learning rate is 0.01. We trained both the networks for 6000 iterations with a batch size of 10, and the curves of training loss are shown in Fig. 10.10.

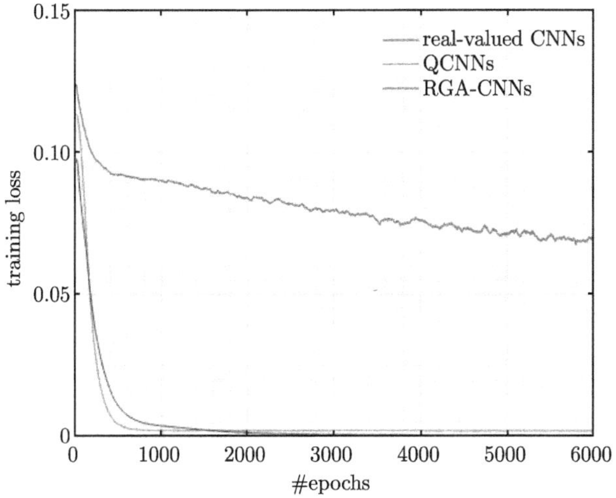

Fig. 10.10 The training loss curves achieved by the real-valued CNNs, QCNNs and the proposed RGA-CNNs

As seen in the training loss in Fig.10.10, we conclude that both QCNNs and the proposed RGA-CNNs converged faster to less training loss than real-valued CNNs. The training loss curves of QCNNs and the proposed RGA-CNNs show stable convergence while real-valued CNNs fluctuates greatly and is very unstable during the training process.

Moreover, less training errors are achieved by QCNNs and the proposed RGA-CNNs compared with real-valued CNNs under the same iterations because of the holistically processing ways provided by quaternion and RGA frameworks. The traning loss curves of QCNNs and the proposed RGA-CNNs exhibit that the convergence and stability are almostly similar. With the increasing number of training iterations from about 200 to 1200, the corresponding training errors of QCNNs are a bit lower than the proposed RGA-CNNs. However, after 1200 iterations, the proposed RGA-CNNs achieve much lower training errors coampared to QCNNs.

For a more objective comparison, we list the test accuracy for CIFAR-10 color image datasets with real-valued CNNs, QCNNs and the proposed RGA-CNNs, in which the test accuracies of 4 classes are presented in Table 10.3. As expected, QCNNs and the proposed RGA-CNNs outperform real-valued CNNs in classification accuracy. As indicated in Table 10.3, the most high test accuracy of real-valued CNNs reaches 80.6% with respect to the third class while that of the proposed RGA-CNNs goes to 90.5% also with regard to the third class. More generally, the total test accuracy obtained by RGA-CNNs is approximately 88.9% while that of real-valued CNNs is only 74.3%. Moreover, it should be noted that the proposed RGA-CNNs shows a bit higher test accuracy than QCNNs, though the test accuracy of the first class is 0.1% lower. That is to say, the experimental results show that the proposed method achieves nearly the highest test accuracy, which verify the advantages of the proposed RGA-based CNNs.

Table 10.3 The test accuracy of the real-valued CNNs, QCNNs and the proposed RGA-CNNs (%)

methods	the first class	the second class	the third class	the fourth class	total
real-valued CNNs[10]	75.9	76.4	80.6	77.3	77.51
QCNNs[14]	85.8	87.0	88.3	88.4	87.4
RGA-CNNs	85.7	88.4	90.5	89.0	88.9

Particularly, with the utilization of RGA framework, the proposed RGA-CNNs method can achieve much better performance, that is to say lower training loss and higher test accuracy, compared with the traditional real networks. This is because the proposed RGA-CNNs method has the good learning ability and explores the inherent characteristics existed in the color images, which play important roles in color image classification tasks.

10.5.5 Complexity Analysis

In this section, we will evaluate the efficiency of QCNNs and the proposed RGA-CNNs with respect to the computation time. The approximate computation time is calculated by using MATLAB on a computer with Intel® Core™ i5-3470 3.20GHz CPU and 4GB memory. The results are shown in Table 10.4 with the computation time of QCNNs and the proposed RGA-CNNs based on 3D geometrical shapes and color images databsets. As shown in Table 8.4, the proposed RGA-CNNs efficiently achieves less computation time, which is nearly the half of QCNNs, for the reason that the RGA framework provides a powerful way to represent a color image as a multivector and process it in a holistic manner. In this way, the relationship of different channels in color image can be completely preserved and the network has been simplified due to the communicative multiplication of RGA.

Table 10.4 Computation time (s) of QCNNs and the proposed RGA-CNNs

datasets	QCNNs	RGA-CNNs
3D geometrical shapes	2772.5113	1678.2767
color images	3144.7837	1681.9981

10.6 Conclusions

In this chapter, a reduced geometric algebra (RGA)-based convolutional neural networks (CNNs) have been proposed, namely RGA-CNNs, which successfully extended the traditional real-valued CNNs into RGA spaces, and investigated its feasibility and characteristics in the application of classification tasks. Taking advantage of the RGA framework with commutative multiplication and none real part, which is bound to remove the data redundancy and simplify the structure of the networks extremely, the proposed RGA-CNNs process the multi-dimensional data in a holistic manner without losing the relationship of different dimensions. Simultaneously, the convolution, pooling, hidden and output layers also have redefinitions based on the RGA theory in detail for the accomplishment of the RGA networks. To demonstrate the effectiveness of the proposed RGA-CNNs, 3D geometric shapes database and color images database are utilized to perform classification experiments, and the classification results are compared with those of real-valued CNNs and QCNNs. Specifically, the proposed RGA-CNNs indeed show optimal performance with lower training loss, higher test accuracy and faster convergence rate in contrast to real-valued CNNs and less computation time compared to QCNNs. Although the proposed RGA-CNNs achieve different performance with respect to the two different databases, the superiority is also prominent in comparison to real-valued CNNs and QCNNs. It is expected that the proposed RGA-CNNs will be more useful in multi-dimensional data processing.

References

[1] Ross G, Jeff D, Trevor D, et al. Rich feature hierarchies for accurate object detection and semantic segmentation. CVPR′14. Cham: Springer, 2014: 2-9.

[2] Yaniv T, Ming Y, Marc A R, et al. Closing the Gap to Human-Level Performance in Face Verification. Conference on Computer Vision and Pattern Recognition (CVPR), 2014: 8.

[3] Krizhevsky A, Sutskever L, Hinton G E. Imagenet classification with deep convolutional neural networks. Advances in Neural Information Processing Systems 25. Curran Associates, Inc., 2012, 11(1): 1097-1105.

[4] Dong C, Loy C C, He K, et al. Learning a deep convolutional network for image super-resolution. European Conf. Computer Vision, 2014.

[5] Wen Z S, Jose C, Ferenc H, et al. Real-time single image and video super-resolution using an efficient sub-pixel convolutional neural network. Computer Vision and Pattern Recognition, 2016, 12: 27-30.

[6] Dong C, Chen C L, Tang X. Accelerating the super-resolution convolutional neural network. European Conf. Computer Vision, 2016: 391-407.

[7] Lefkimmiatis S. Non-local color image denoising with convolutional neural networks. arXiv, 2016.

[8] Lawrence S, Giles C L, Tsoi A C, et al. Face recognition: A convolutional neural-network approach. IEEE Transactions on Neural Networks, 1997, 8(1): 98-113.

[9] Nebauer C. Evaluation of convolutional neural networks for visual recognition. IEEE Trans Neural Netw, 1998, 9(4): 685-696.

[10] Zhang F, Cai N, Wu J, et al. Image denoising method based on a deep convolution neural network. IET Image Processing. IEEE, 2018, 12(4): 485-493.

[11] Li Q, Peng Q, Yan C. Multiple VLAD encoding of CNNs for image classification. Computing in Science Engineering, 2018, 99: 1.

[12] Zhang X, Zou J, He K, et al. Accelerating very deep convolutional networks for classification and detection. IEEE Transactions on Pattern Analysis Machine Intelligence, 2015, 38(10): 1943-1955.

[13] Guberman N. On Complex Valued Convolutional Neural Networks, 2016.

[14] Kominami Y, Ogawa H, Murase K. Convolutional neural networks with multi-valued neurons. IEEE Transactions on Pattern Analysis and Machine Intelligence, 2010, 32(9): 1582-1596.

[15] Shen M M, Wang R, Cao W M. Joint sparse representation model for multi-channel image based on reduced geometric algebra. IEEE Access. Image Process, 2018, 6:

24213-24223.

[16] Gunn C G. A new approach to euclidean plane geometry based on projective geometric algebra. Mathematics, 2015, 704(4): 41-58.

[17] Wareham R J, Cameron J, Lasenby J. Applications of conformal geometric algebra in computer vision and graphics. IWMM2004, GIAE2004, LNCS, 2004, 3519: 329-349.

[18] Lopes W B, Al-Nuaimi A, Lopes C G. Geometric-algebra LMS adaptive filter and its application to rotation estimation. IEEE Signal Processing Letters, 2016, 23(6): 858-862.

[19] Ebling J, Scheuermann G. Clifford Fourier transform on vector fields. IEEE Transactions on Visualization Computer Graphics, 2005, 11(4): 469-479.

[20] Wang R, Zhou Y X, Jin Y L, et al. Sparse fast Clifford Fourier transform. Frontiers of Information Technology Electronic Engineering, 2017, 18(8): 1131-1141.

[21] Li Y, Liu W, Li X, et al. GA-SIFT: A new scale invariant feature transform for multispectral image using geometric algebra. Information Sciences, 2014: 559-572.

[22] Li H B, Cao Y. On geometric theorem proving with null geometric algebra. Guide to Geometric Algebra in Practice, 2011: 195-215.

[23] Li H B. Symbolic Geometric Reasonong with Advanced Invariant Algebra. International Conference on Mathematical Aspects of Computer and Information Sciences, 2015: 35-49.

Color Figures

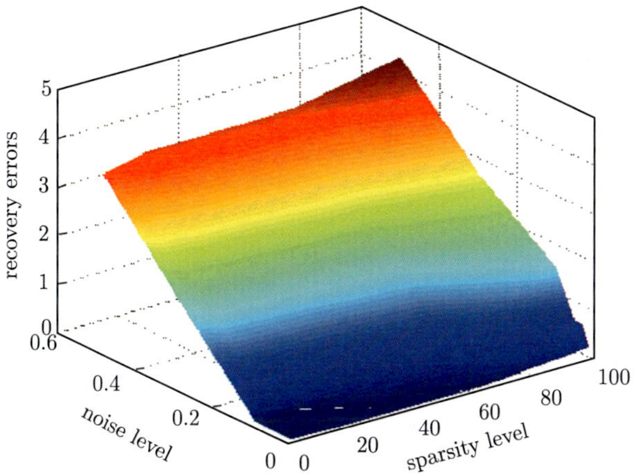

Fig.1.5　Averaged recovery errors of 8-dimensional signals as a function of noise level σ and sparsity level t

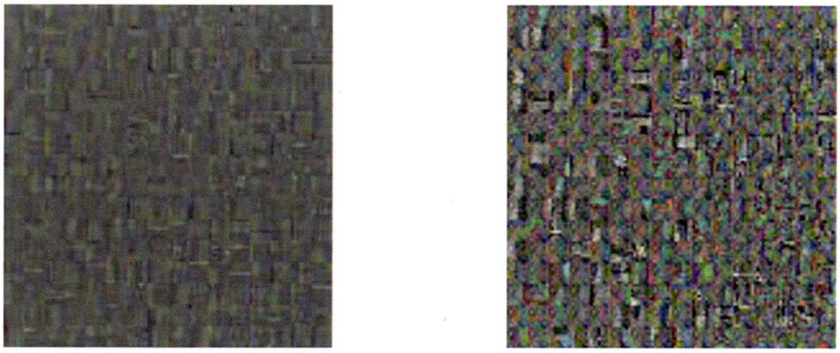

(a) K-SVD　　(b) K-GASVD

Fig.3.1　Visualization of the color image dictionaries learned by K-SVD (a) and K-GASVD (b) methods

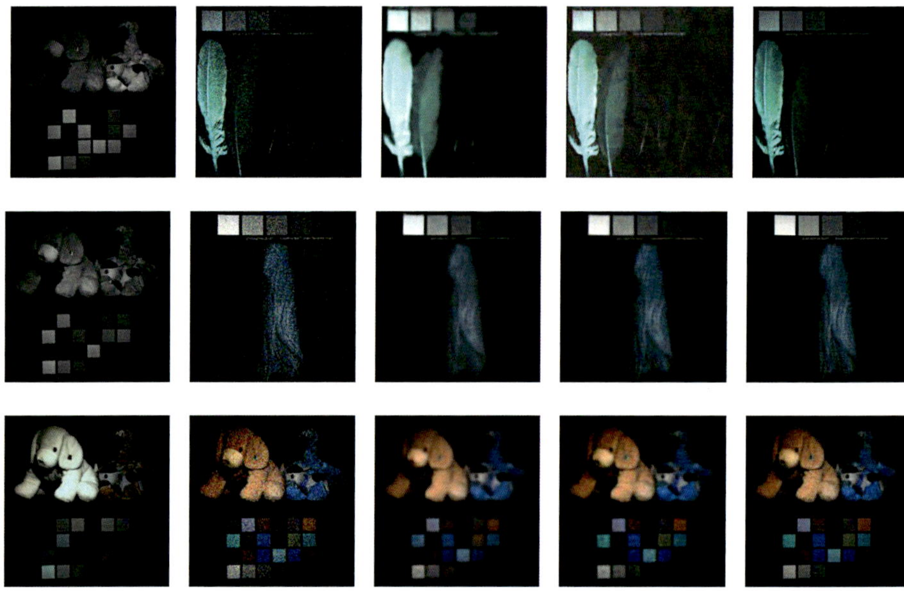

(a) Original images (b) Noisy images ($\delta = 30$) (c) K-SVD (d) Improved K-SVD (e) K-GASVD

Fig.3.6 Denoising results of multispectral images

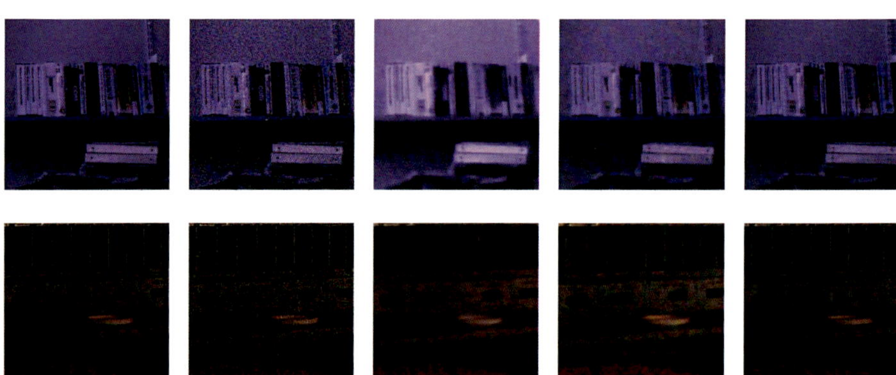

(a) Oiginal images (b) Noisy images ($\delta = 30$) (c) K-SVD (d) Improved K-SVD (e) K-GASVD

Fig.3.7 Denoising results of hyperspectral images

(a) GA-SURF

(b) GA-SIFT

Fig. 4.4 Interest points extracted from multispectral image ($n = 31$)

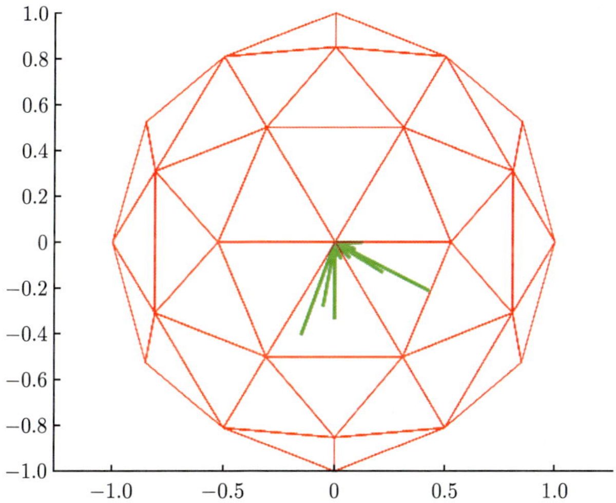

Fig. 6.7 80 gradient directions. The whole gradient histogram contains 80 bins

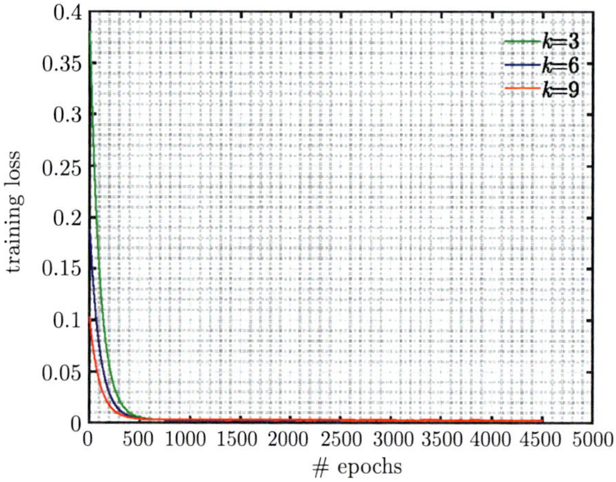

Fig. 7.3 The training loss on GA-CNNs with different number of convolutional kernels

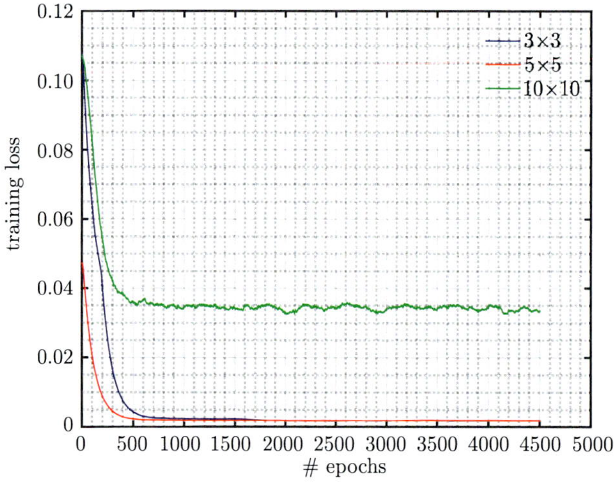

Fig. 7.5 The training loss on GA-CNNs with different pooling size

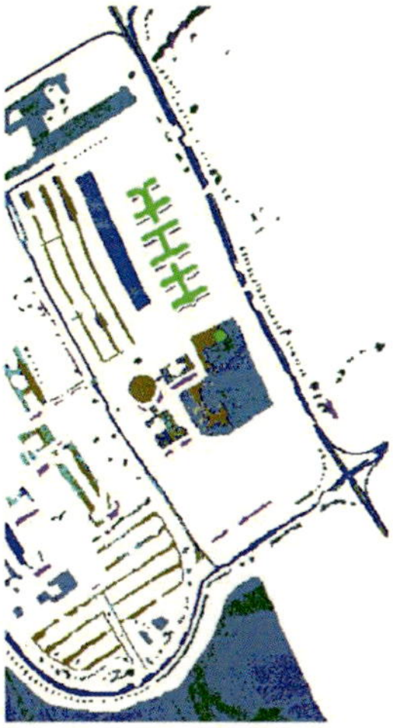

(a) Classification results by real-valued GA-CNNs with background

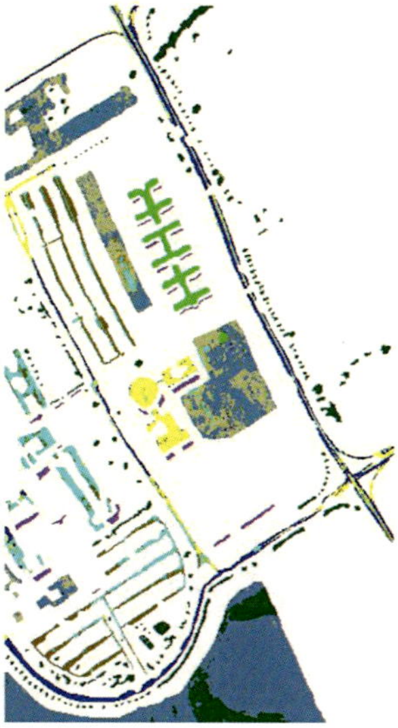

(b) Classification results by real-valued without background

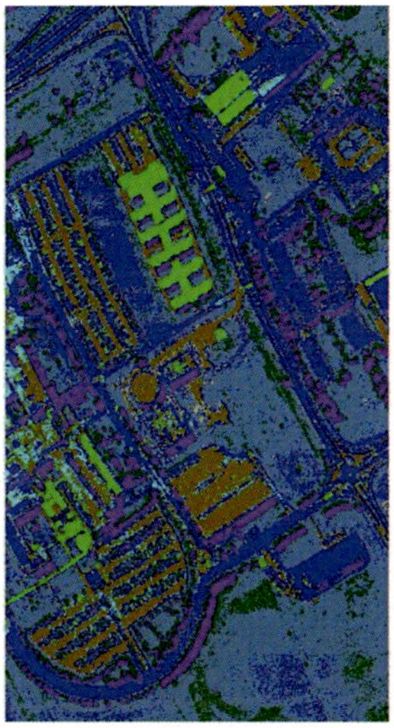

(c) Classification results by real-valued CNNs model with background

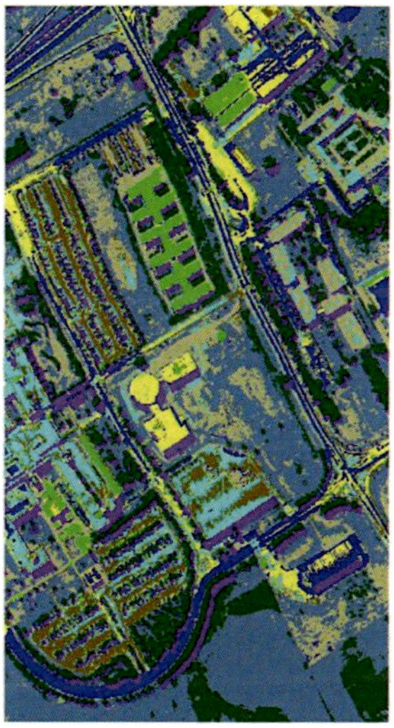

(d) Classification results by GA-CNNs model with background

Fig. 7.9 Classification results for Pavia U dataset

(a) Original images (b) Noisy images (σ=20) (c) K-SVD (d) K-QSVD (e) K-RGASVD

Fig. 8.5 Denoising results

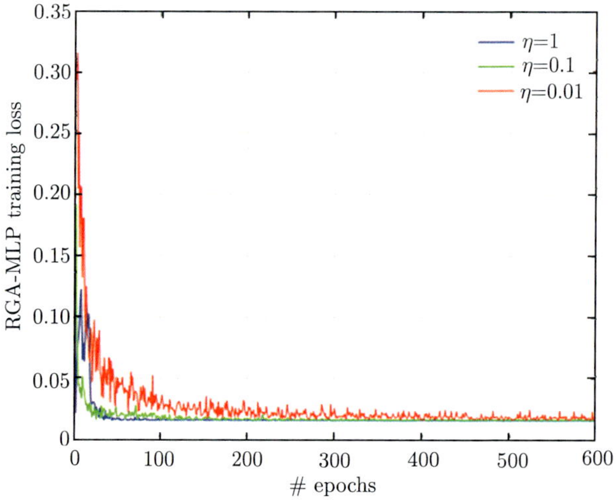

Fig. 9.4 The training loss curves achieved by the proposed RGA-MLP model with different learning rate η

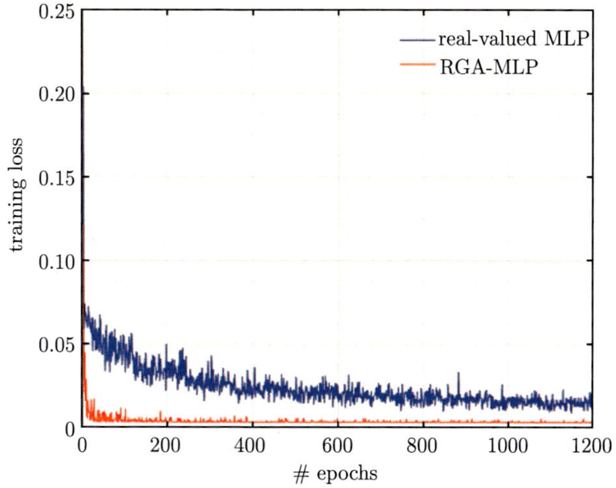

Fig. 9.5 The training loss curves achieved by the real-valued MLP and the proposed RGA-MLP models

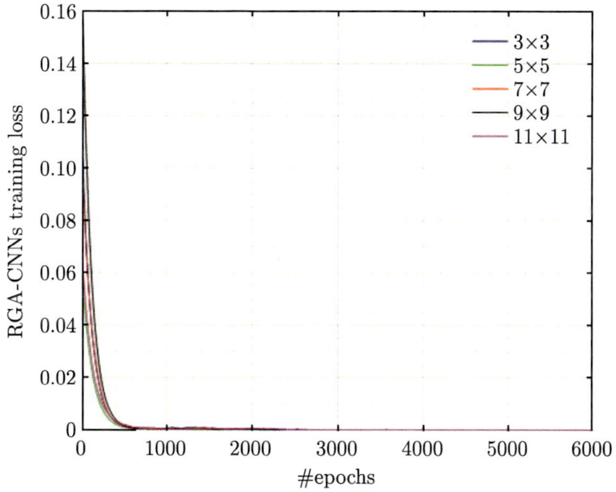

Fig. 10.7 The training loss curves achieved by the proposed RGA-CNNs with different size of convolution kernels

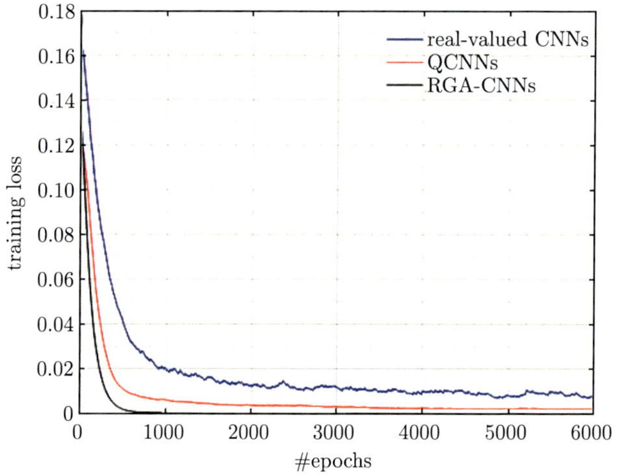

Fig. 10.8 The training loss curves achieved by the real-valued CNNs, QCNNs and the proposed RGA-CNNs

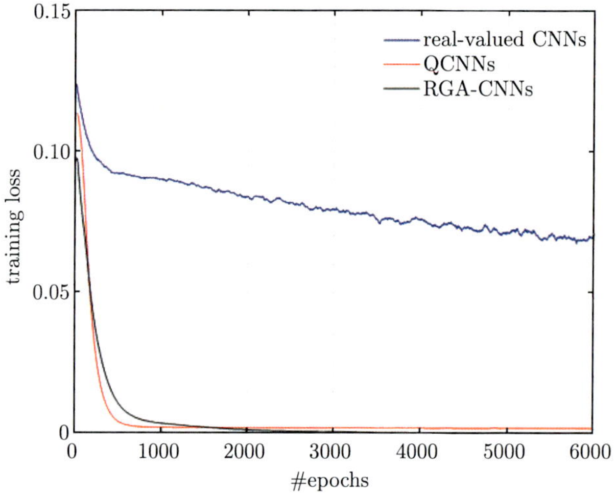

Fig. 10.10 The training loss curves achieved by the real-valued CNNs, QCNNs and the proposed RGA-CNNs